Microcomputer Knowledge-Based Expert Systems in Civil Engineering

Proceedings of a symposium sponsored by the
Structural Division of the
American Society of Civil Engineers
in conjunction with the
ASCE National Convention, Nashville, Tennessee

May 10-11, 1988

Edited by Hojjat Adeli

Published by the
American Society of Civil Engineers
345 East 47th Street
New York, New York 10017-2398

ABSTRACT

This book contains sixteen papers presenting applications of expert system technology to civil engineering problems with emphasis on microcomputer implementations. It is divided into four parts: Structural engineering, geotechnical and environmental engineering, construction, and general. Topics include knowledge acquisition and machine learning, using PROLOG on a Macintosh, an environment for building integrated structural design expert systems, an integrated rule-based system for industrial building design, and integrating an expert system shell with spreadsheet programs. Expert systems for hazardous waste management, diagnosis and treatment of dam seepage problems, and analysis of activated sludge are presented. Also covered are knowledge elicitation techniques for construction scheduling, an expert system for construction contract claims, and knowledge acquisition for a contractor prequalification knowledge-based system. Finally, logic programming to manage constraint-based design, and development of an earthquake insurance and investment risk analysis system are discussed.

Library of Congress Cataloging-in-Publication Data

Microcomputer knowledge-based expert systems in civil engineering: proceedings of a symposium sponsored by the Structural Division of the American Society of Civil Engineers in conjunction with the ASCE National Convention, Nashville, Tennessee, May 10-11, 1988; edited by Hojjat Adeli.
    p. cm.
  Includes indexes.
  ISBN 0-87262-653-9
    1. Civil engineering—Data processing—Congresses. 2. Expert systems (Computer science)—Congresses. 3. Microcomputers—Congresses. I. Adeli, Hojjat. II. American Society of Civil Engineers. Structural Division.
TA345.M486 1988   88-7041
624'.028'5633—dc19   CIP

The Society is not responsible for any statements made or opinions expressed in its publications.

Authorization to photocopy material for internal or personal use under circumstances not falling within the fair use provisions of the Copyright Act is granted by ASCE to libraries and other users registered with the Copyright Clearance Center (CCC) Transactional Reporting Service, provided that the base fee of $1.00 per article plus $.15 per page is paid directly to CCC, 27 Congress Street, Salem, MA 01970. The identification for ASCE Books is 0-87262/88. $1 + .15. Requests for special permission or bulk copying should be addressed to Reprints/Permissions Department.

Copyright © 1988 by the American Society of Civil Engineers,
All Rights Reserved.
Library of Congress Catalog Card No.: 88-7041
ISBN 0-87262-653-9
Manufactured in the United States of America.

PREFACE

This book contains the papers accepted for the sessions on "Microcomputer Knowledge-Based Expert Systems in Civil Engineering" organized by the editor in Nashville, Tennessee, May 10-11, 1988. In response to calls for papers, twenty nine abstracts were received. Twenty one papers were tentatively selected on the basis of details presented in the 500-word abstracts. Authors of sixteen papers produced acceptable papers in due time. These papers are included in this book. All papers are eligible for discussion in the appropriate journal. All papers are eligible for ASCE awards.

Hojjat Adeli
Editor

CONTENTS

Part I Structural Engineering 1

Knowledge Acquisition and Machine Learning in an Expert System
  X.J. Zhang and J.T.P. Yao 2
Expert System RAISE-1
  R. Chen and X. Liu 16
Using PROLOG on a Macintosh to Build an Engineering Expert System
  W.M. Kim Roddis and J. Connor 26
SDL: An Environment for Building Integrated Structural Design Expert Systems
  Y. Paek and H. Adeli 40
An Integrated Rule-Based System for Industrial Building Design
  B. Kumar and B.H.V. Topping 53
A Rule-Based System for Estimating Snow Loads on Roofs
  P. Fazio, C. Bedard, and K. Gowri 73
PC PLUS + LOTUS 123
  S. Malasri 79
A Knowledge-Based Expert System for the Selection of Structural Systems for Tall Buildings
  P. Jayachandran and N. Tsapatsaris 88

Part II Geotechnical and Environmental Engineering 100

GEOTOX-PC: A New Hazardous Waste Management Tool
  G.K. Mikroudis and H.-Y. Fang 101
An Expert System for Diagnosis and Treatment of Dam Seepage Problems
  M.I. Asgian, K. Arulmoli, W.O. Miller, and K. Sanjeevan 117
ASA: An Expert System for Activated Sludge Analysis
  D.G. Parker and S.C. Parker 126

Part III Construction 138

Knowledge Elicitation Techniques for Construction Scheduling
  J.M. De La Garza, C.W. Ibbs, and E.W. East 139
An Expert System for Construction Contract Claims
  M.P. Kim and K. Adams 153
Knowledge Engineering in a Knowledge-Based System for Contractor Prequalification
  J.S. Russell and M.J. Skibniewski 168

Part IV General 185

   Logic Programming to Manage Constraint-Based Design
     W. Chan and B.C. Paulson, Jr. 186
   Expert Systems for the Earthquake-Related Industry
     J. Kim, W. Dong, F. Wong, and H.C. Shah 201

Author Index 210

Subject Index 211

PART I

STRUCTURAL ENGINEERING

KNOWLEDGE ACQUISITION AND MACHINE LEARNING IN AN EXPERT SYSTEM

X.J.Zhang[1], and J.T.P.Yao[1], F.ASCE

1. INTRODUCTION

Damage assessment and safety evaluation for existing structures pose a challenging problem for structural engineers. Many difficulties of assessment and evaluation for an existing structure result from the complexities of the structural system and uncertainties of the past and future environmental conditions. Because there are many uncertainties involved during design, construction and operation, the exact behavior of a given structure is still difficult to understand. As a result, it is not yet possible to find a general mathematical model for the precise description of the structural behavior under various environmental and loading conditions. The current procedures of assessment and evaluation of existing structures are highly dependent on the experience, intuition and judgement of recognized experts. Much research work has been focused on finding a rational and systematic process for the assessment of structural damage and evaluation of structural safety in practical situations. Recently, knowledge-based expert system has become a feasible and potentially useful tool in the development of such a rational and systematic assessment process.

SPERIL systems have been under development for damage assessment of existing structures. The name of these systems is taken from Structural PERIL. The first version SPERIL-1 is a rule-based expert system (e.g., Ishizuka, Fu and Yao, 1981). The organization and rules of SPERIL-1 are basically designed to assess earthquake damage of existing building structures. In SPERIL-1, an arbitrary damage measure ranging from zero ("no damage") through ten ("total collapse") is used. Separate evidential observations are integrated with the combined use of Dempster & Shafer's theory and fuzzy sets. The program portion of SPERIL-1 is written in C. SPERIL-2 was developed (e.g., Ogawa, Fu and Yao, 1985) to assess the general damage of existing structures. In SPERIL-2, the integer exponent (the order of magnitude) n of failure probability 10^{-n} is used as a measure of structural safety. SPERIL-2 is basically a rule-based system, where logic is used (a) to represent rules, facts and available data on the existing structure and (b) to control the inference to obtain conclusion. The program of SPERIL-2 is written in Prolog. A new version, SPERIL-3, is being developed (Zhang and Yao,1986b) on the basis of a preliminary version called CES-1 (Zhang, 1985), which consists of the following four parts: inference machine, knowledge base, memory and learning machine. A frame structure is used to represent knowledge in the SPERIL-3, and the program is written in Lisp.

---
[1] School of Civil Engineering, Purdue University, W. Lafayette, IN 47907-3399.

The development of the SPERIL-3 system is briefly described. Specifically, knowledge acquisition and machine learning in this system are emphasized herein.

2. DAMAGE ASSESSMENT

The development of previous versions of SPERIL systems has demonstrated the feasibility and potential of practical application of knowledge-based expert systems in damage assessment. The motivations for the development of such a practical knowledge-based expert system for damage assessment are described as follows:

1) Large Domain

Structural assessment is a large domain, which can be characterized as a sequence of decision-making problems. Generally speaking, three tasks are involved during an assessment. The first task is to obtain a diagnosis for possible causes of the damage. The second one is a qualitative evaluation including (a) classifying the current damage state and potential hazards and (b) evaluating the safety (or reliability) for the structures for future usage. Based on results of the diagnosis (first task) and evaluation (second task), a final decision related to maintenance can then be made (third task). Although experts generally agree that a comprehensive assessment can be made using different information sources, such as visual inspection, documentation review, testing and analysis, there exists no standard procedure for summarizing and interpreting those information sources. Moreover, knowledge in the damage assessment domain is not well codified. To-date knowledge acquisition has been a bottleneck in the development of a practical knowledge-based expert system in such a large domain as damage assessment, which is not only time-consuming but also difficult in communicating among experts.

2) Hybrid Tasks

During an assessment process, many tasks are involved and the resulting information should be processed by using various methods. Some information may be treated efficiently by using empirical knowledge such as experience and engineering judgement, while others can be processed more efficiently by using numerical approximation and calculation such as system identification and finite element techniques. The methodologies for damage assessment have been reviewed by Zhang and Yao (1986a). Most methods in damage assessment are not sufficient by themselves because the information may be inexact and/or incomplete. Therefore, the results obtained by using different methods should be combined and compared with one another. The choice of several possible methods also requires expert opinion. Moreover, the final results must be properly interpreted according engineering expertise and judgement (WJE Report, 1987).

3) Multiple Damage Events

The problem solving in most existing diagnosis expert systems is associated with the best solution from a finite list of prespecified answers (hypotheses). In such a diagnosis, only one hypothesis with the highest belief is chosen as the best solution. However, the structural evaluation is a multiple damage events problem. In the diagnosis for damage causes, one or more hypotheses can be chosen at the same time. There are two issues in a problem with multiple hypotheses. One is how to make the inference for diagnosis of multiple damage events. The other is how to combine multiple damage events for evaluating the reliability of the structure.

4) Various Types of Uncertainties

There exist various types of uncertainties in damage assessment. Brown and Yao (1983) classified engineering problems into four different type according to the type of uncertainties. To deal with different types of uncertainties, different approximate reasoning methods are needed. Decision-making in damage assessment and safety evaluation is a subjective and context-dependent process. Several approximate reasoning methods in SPERIL systems are discussed by Zhang and Yao (1986b). More details about inference methods for dealing with uncertainties in damage assessment problem along with decision in multiple damage events in the SPERIL-3 will be described in another paper in the near future.

5) Heuristic Strategies

Many heuristic strategies can be applied during a damage assessment process. To understand the decision-making process and to acquire heuristic strategies of an expert are difficult tasks in knowledge acquisition for the development of any expert system. Heuristic strategies applied in the SPERIL systems have been discussed (Zhang and Yao, 1986b). A heuristic strategy called the alpha-beta procedure is also applied in the SPERIL-3 for manage information processing (Zhang and Yao, 1987).

3. KNOWLEDGE ORGANIZATION

Subjective knowledge and experience of an expert imply a higher level of intellectual organization. To transfer this kind of knowledge into expert system is a higher level of intellectual task, which is not a mechanical copy but a process of learning and comprehension of the domain knowledge. It is estimated that a knowledge engineer may spend more than ninety percent of time in understanding, formalizing and organizing the relevant knowledge while he/she develops an expert system (eg., Cheng, 1986). Therefore, it is important to emphasize (a) the learning of the overall organization of knowledge, and (b) the acquisition of control knowledge about when and how to use (or not to use) certain facts. A good knowledge organization is desirable in an expert system for easy to acquire knowledge and efficient to solve the problem.

Because many tasks are involved and influenced each other, a simple tree structure is difficult to organize the knowledge of this domain. From a psychological viewpoint, the human memory may consist of clusters of symbols called "chunks", which refer to symbols associated with a set or pattern of stimuli. Logical chunks are grouped together in long term memory by their common associations, and by physically separating these chunks from others (Laird et. al. 1986). Learning and remembering occur as linkages between chunks are established and revised. A ´chunking model´ is used to model the organization of expert knowledge (Scott et. al., 1986) and then developed for knowledge acquisition and machine learning in the SPERIL-3 system.

3.1 Problem Decomposition

There are the following two ways to decompose an assessment problem, namely damage state decomposition and information source decomposition. In the former, a damage state is decomposed into several damage events (or causes of damage) which may further be decomposed into even simpler events. For example, a damage state can be decomposed into damage events such as corrosion, overstress, fatigue, fracture, and creep. In the latter, the assessment problem is decomposed by the information sources which are combined for solving the problem. For instance, the information source is decomposed into such as general information, visual inspection, field survey, document review, field tests, laboratory tests, earthquake records, analysis and othes. In the SPERIL-3, an assessment problem is decomposed into two separate spaces, namely events space (problem space) and task space (information space), which are related by their common associations.

3.2 Overall Organization of Knowledge

As suggested by Yao et al. (1984), an assessment problem may be solved by using an iterative procedure. As engineers obtain relevant information and test data, analysis and evaluation are performed. If and when results are sufficient for determining the structural condition, the process is complete. Otherwise, more inspection information and test data must be collected for further analysis and evaluation. The process is repeated until the structural condition is assessed with some degree of confidence. In a multiple damage events problem, an assessment process may be interpreted by the following steps:

1) to establish a set of suspected damage events from event space according to preliminary information;

2) to suggest investigations from task space appropriate for the suspected damage events;

3) to analyses, diagnose, evaluate and summarize each suspected damage event by using an iterative procedure;

4) to combine all possible damage events;

5) If the combined results are sufficient for interpreting the structural condition, then to recommed possible structural repair method and maintenance strategies; Otherwise, go to step 1 and repeat steps 1 through 4 until satisfactory results are obtained.

The overall control flow of the SPERIL-3 for the assessment and evaluation of multiple damage events in a structure is shown as Fig.1.

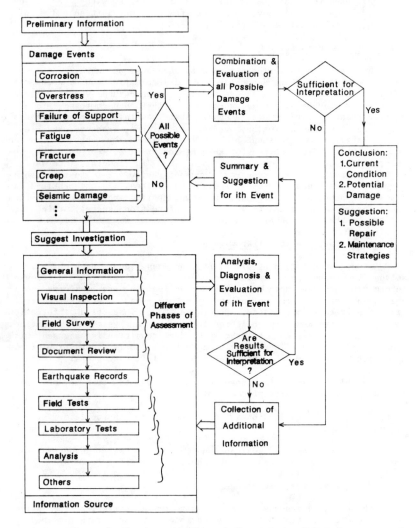

Figure 1 The Overall Control Flow of the SPERIL-3

4. KNOWLEDGE REPRESENTATION

4.1 Frame

A frame structure is used to represent knowledge in the SPERIL-3. In the frame structure, knowledge is represented as a structured object. A frame usually consists of a frame name and several slots. One slot has only one name and may have several values. A slot value can be a constant, a variable or a function. As an illustration, a frame about the half-cell field testing method is shown in the figure 2. In this frame, the title "Half-cell Test" is called frame name. Such construction as "AKO: Field Testing Method" is called a slot. The first expression, "AKO", is called the "slot name", and second expression, "Field Testing Method" is called the "slot value". The slots "AKO" indicates the classification of the frame; "DEF" gives the definition of the half-cell test method; "PCOND" specifies the pre-conditions of using the half-cell testing method; "ACT" indicates the actions which should be taken after the preconditions are satisfied in certain degree; and "CRIT" indicates the criteria of test results for the evaluation of corrosion activity.

```
Half-cell Test
    AKO:  Field Testing Method
    DEF:  Half-Cell Potential Measurement for
          Corrosion Activity of Reinforced Concrete
    PCOND: 1) Material: RC
           2) Possible Reason of Damage from Visual Inspection:
              Corrosion
           3) Damage Location: Top Face of Clab
           4) Importance of the Structure:
              At Least Important
           5) The Environment: Wet
     ACT:  1) Ask the Questions Related to Testing Results
           2) Evaluate Test Results
     CRIT:    Test Results      Evaluation       Certainty
              V < -0.2 volts    not active       0.9
              V > -0.35 volts   active           0.9
              -0.2 to -0.35     Unknown
```

Figure 2   The Frame: Half-Cell Test

In the frame-like structure, each slot can also be described with another frame. For example, the material may be described by another frame. The relevant information about the classification of the structural material may be contained in corresponding slots. In this manner, complex knowledge may be represented by using a series of frames. It is convenient to represent sequences of events by using a frame-like structure. One of the main advantages of frame representation is that the representation of information is explicit. Because each slot in a frame has its own name and values, which are explicit. By checking each node corresponding to each slot value, the available information can be obtained or the missing information will be requested.

4.2 Control Knowledge

In the SPERIL-3, the information processing is controlled in two ways. One is that the information is controlled by the preassigned order as shown in Fig.1, which is corresponding to different phases of assessment in practice. Another way is that the effective control procedure is performed by employing such slots as AKO, ISA, CAUSE, REASON, PCOND and ACT. In a frame, the slots PCOND and ACT play a similar role as "IF-THEN" rule. If the preconditions are satisfied in a certain degree, then the actions will take place. Each precondition may correspond to a question-answers or another frame.

The function "derive" in the SPERIL-3 is used to derive a frame according to the preconditions. A frame can be derived in two phases, namely the elaboration phase and the decision phase. The purpose of the first phase of the derivation is to check the possibility of the derivation of a frame. If and only if it is possible to derive the frame, the second phase of the derivation (the decision phase) will be followed. There are two cases when it is not possible to derive a frame. One is that some pieces of available information in the working memory have not satisfied the preconditions. In this case, the second phase of the derivation will be skipped and no question will be asked. In the second case, some pieces of information in the knowledge base are missing. In this case, a warning about the missing information will be issued and the machine prepares to learn. The control flows for two phases of the derivation of a frame are shown in Fig.3.

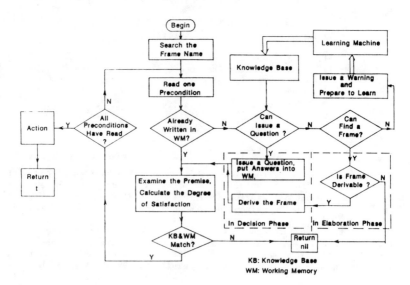

Figure 3 Derive One Frame

5. MACHINE LEARNING

5.1 General Remarks

There is an evolutionary process inherent in the process of the creation, testing and validation of any knowledge-based system. For a system containing vast amounts of knowledge, the knowledge base must be developed gradually and incrementally. The system needs much testing, extending and improving on a continuous basis. No rational body of knowledge is ever complete and unchanging. Therefore, it is desirable to incorporate a learning machine into a knowledge-based expert system. For example, by incorporating TEIRESCAS (Davis, 1977), MYCIN can be used to acquire knowledge from experts interactively. A learning machine called "LEARN" is incorporated into the SPERIL-3 system. The learning machine is used herein to mainly accomplish the following three tasks: (a) to acquire new knowledge, (b) to verify the existing knowledge base, and (c) to modify the knowledge base. Learning may be classified into two basic forms, i.e., knowledge acquisition and skill refinement. The former means learning new symbolic information and the ability to apply this information effectively. The latter is the gradual improvement of motor and cognitive skill through practice (Carbonell, et. al., 1983). The discussion of this paper is concentrated on one of the essential problems in the machine learning, the automation of knowledge acquisition and knowledge organization.

The automatical knowledge acquisition in a learning machine can be designed to facilitate communication with experts and users in English about domain knowledge as described by Zhang and Yao (1986b). The knowledge acquisition procedure may be implemented by adding, deleting and/or modifying slots and corresponding values in frame structures. When a new concept is introduced to machine and a new subset is created in the knowledge base of the system. The relationships among the new concept and existing knowledge may be established through the experts and machine interface. In any event, it is necessary for the expert to constantly monitor and evaluate the additional knowledge in order to decide that such modifications represent an improvement of the system as intended. On the other hand, it is necessary and important to check the relationship such as redundancy or inconsistency between the new knowledge and existing knowledge. In this manner, an automatical knowledge acquisition procedure may be followed by three steps: a) add new knowledge into the knowledge base, b) check the redundancy or inconsistency between new knowledge and existing knowledge, and c) modify and reorganize the knowledge base. The first step of knowledge acquisition may be considered as learning through the instructor (Winston, 1984). However, the last two steps, the machine should "think" by itself and will help the instructor to carefully check, modify and reorganize the knowledge base.

5.2 Knowledge Acquisition Model

A concept called conditional knowledge is introduced by Cheng (1986). A conditional is a relation between two concepts in the form of "C1 R C2", where C1 and C2 are the concepts and R is a relation. For instance, a proposition P "A is a kind of B" can be represented as "A AKO B". Two kinds of relational terms, namely the classificational relation and the causal relation are used in this study (Zhang & Yao, 1986b). The former may include such as AKO (a kind of), ISA (is a), PARTOF (is a part of) and HASPART (has part), which make class membership and subclass relations explicit, and facilitating the movement of knowledge from one level to another. The latter can be such as CAUSE (cause), REASON (is a reason of), EVIN (is an evidence of) and CONC (is a conclusion of), which link the causaling and resulting events.

A relational term may have its inverse. That means if C_1 R C_2 hold, then C_2 R^{-1} C_1 also hold. The pair of relational terms R and R^{-1} is defined as an inverse pair. Several inverse pairs of relational terms may be defined as the follows:

R	R^{-1}
AKO	ISA
PARTOF	HASPART
CAUSE	CAUSED-BY
CAUSE	REASON
EVIN	CONC

Conditionals may be chained together through the composition of their relations. From conditionals "C1 R1 C2" and "C2 R2 C3", another conditional "C1 (R1*R2) C3" may be implied. Cohen,et.,(1985) gave an example about this inference. If x causes y and z is a kind of x, then z also cause y. The key of the credibility of these inferences depends not only on x, y or z, but also on the understanding of the relational terms such as CAUSE and AKO. For example, from the conditionals "inadequate strength of a structure CAUSE overstress" and "overstress AKO damage", a new conditional "inadequate strength CAUSE damage" can be implied. In this conditional chain, the new relation "R = CAUSE*AKO = CAUSE" is implied, and usually "R = CAUSE*AKO = CAUSE" is unique. However, conditionals are not always uniquely chained together. For instance, from conditionals "corrosion CAUSE crack" and "crack ISA corrosion crack", the new conditional "corrosion CAUSE corrosion crack" can be implied. On the other hand, if the two conditionals are "corrosion CAUSE crack" and "crack ISA shear crack", no new conditional can be implied from chaining those two conditionals. It is evident that the chain "R = CAUSE*ISA" is not uniquely defined relation.

Based on above discussion, several special chains are defined as the follows:

1) "Unique" chain

If a relation S through the composition of two relations R1 and R2 is uniquely defined, then the chain "S = R1*R2" is called as a "Unique" chain. In the "Unique" chain, the relation S definitely holds, no matter what concepts between the relation are. For instance, "S = AKO*AKO = AKO" is a "Unique" chain. If the propositions "A is a kind of B" and "B is a kind of C" hold, then the proposition "A is a kind of C" must hold. In general, if "set A belongs to set B" and "set B belongs to set C" hold, then "set A belongs to set C" also hold. Several examples of unique chains may include S = AKO*AKO = AKO; S = PARTOF*PARTOF = PARTOF; S = CAUSE*CAUSE = CAUSE.

2) "May be" chain

If a relation S through the composition of two relations R1 and R2 can not be uniquely defined, then the chain "S = R1*R2" is called as a "May be" chain. This means that the new conditional may be implied through the composition of two relations R1 and R2, but it is not always true. For instance, the chain "R = CAUSE*ISA" is a "May be" chain. If "A CAUSE B" and "B ISA C" hold, then the proposition "A CAUSE C" may hold depending on concepts A, B and C.

3) "Sibling" chain

If a relation S through the composition of any relation with its inverse, namely the chain "S = R* R^{-1}" is called as a sibling chain. When a "Sibling" relation between two concepts, there could be three different cases: 1) a hidden relation between two concepts; 2) intermediate concepts with a chain between two concepts; 3) no relation between them. Therefore, the search of possible relations between two concepts may be needed.

A chain with n relations "S = R1*R2*...*Rn" can be composited as $_{-1}$S = {((R1*R2)*...*Rn). The inverse of a chain is defined as (R1*R2)$^{-1}$ = R2^{-1}* R1^{-1}. The inverse of a chain will keep the feature, namely the inverse of an "Unique" chain is still a "Unique" chain, so and other two.

5.3 <u>Knowledge Check</u>

Algorithms for checking consistency and redundancy of rule-based systems have been proposed by Nguyen et.al (1985). An algorithm for checking consistency or redundancy by the composition of the conditionals is suggested by Cheng (1986). The method of the knowledge check developed in this paper is based on the composition of the conditionals.

Suppose there already exists a knowledge body KB containing a set of concepts C = { C1, C2,..., Cm }. When a new concept Cnew is fed into the knowledge base, the new concept and the relationship among it and others can be defined through the user and computer interface. The redundancy or inconsistency check between new knowledge and existing knowledge becomes a necessary and important task during the process of knowledge acquisition. Without loss generality, suppose two relations are defined as Rnew between C1 and Cnew, and Rnew+1 between Cnew and Ck. The general criteria of knowledge check may be described as the follows:

1) If there is no chain between C1 and Ck before the new knowledge, then the new knowledge is accepted without change.

2) If there is a chain between C1 and Ck, say R1, R2 .... Rk-1, then check this chain with new relations Rnew and Rnew+1. For more than one chain between C1 and Ck, check them with new relations separately.

3) If R = Rnew*R1*R2....*Rk-1 is a "Unique" chain and the relations R is equal to Rnew+1, then the new relations is redundant with the existing knowledge.

4) If R = Rnew*R1*R2....*Rk-1 is a "May be" chain, then the new knowledge may be accepted after knowledge confirmation.

5) If R = Rnew*R1*R2....*Rk-1 is a "Sibling" chain, then check R = $Rnew^{-1}$*Rnew+1.

6) If R = Rnew*R1*R2....*Rk-1 is inconsistent with Rnew+1, then the new relations is inconsistent with the existing knowledge.

After knowledge check three kind of actions may be taken: 1) new knowledge is accepted without change in case 1 as satisfying the criterion 1, 2) knowledge modification or reorganization either in case 3, 5 or 6, 3) knowledge confirmation in case 4.

5.4 Triangle Model for Knowledge Reorganization

The automation of knowledge reorganization in a knowledge acquisition process discussed in this paper is concentrated on the case 3 and 5 as mentioned in knowledge check. Usually, case 3 and case 5 may happen together, in which S1 = Rnew*R1*R2....*Rk-1 is a "Sibling" chain, and S = $Rnew^{-1}$*Rnew+1 is a "Unique" chain and equals to S1 = Rnew*R1*R2....*Rk-1.

A triangle model is developed here for knowledge reorganization. A triangle structure of semantic network indicated three concepts with three relations is shown as Fig.4a which indicates three conditionals "C1 R1 C2", "C3 R2 C1" and "C3 R3 C2". Suppose it is known that the chain "S = R2*R1" is a "Sibling", and S1 = $R2^{-1}$*R3 is redundant with R1. A simple rule for knowledge reorganization of this triangle network can be described as the following two steps: 1) cancel the relation $R1_{i}$, and 2) replace the conditional " C3 R2 C1" by the conditional " C1 $R2^{-1}$ C3 ". A string structure of network is shown in Fig.4b after knowledge reorganization, which indicates two conditionals: "C1 $R2^{-1}$ C3" and "C3 R3 C2". No redundant exists in the new string structure of network and the relation between C1 and C2 is implied uniquely.

The suggested triangle model can be extented to a general case. A semantic network of k concepts with k-1 relations and a new concept Cnew as described in knowledge check. If the relationships between new concept Cnew and each one of the existing concepts in the set { C1,...,Ck } are established, at most k triangle structures may be created. By using triangle model step by step this semantic network may be reorganized.

A simple example in a classification network is given here as an illustration of the knowledge reorganization. Suppose the knowledge acquisition is to learn some concepts related to structural material. Four concepts, say RC (reinforcing concrete), CON (concrete), PRC (prestressed reinforcing concrete) and ST (steel), are introduced by the instructor. The relationships between the concept "material" and those four concepts are also established as "RC AKO material", "CON AKO material", "PRC AKO material" and "ST AKO material". It is evident that six sibling relations have been implied among the four concepts including sibling (RC, CON), sibling (CON, PRC), and so on. Because of the "Sibling" relationships among the four concepts it is needed to search other possible relationships among them. During the search three conditionals are acquired through the instructor, namely "RC AKO CON," "CON ISA PRC" and "RC ISA PRC". The semantic network of the relationships among the concepts are shown in Fig.4c. Three triangles of semantic network are created among the four concepts material, RC, CON and PRC. By using the triangle model separately, the relation AKO between material and PRC, the relation AKO between material and RC, and the relation ISA between CON and PRC are canceled. The relation ISA between RC and CON is replaced by the relation "AKO" (the inverse of "ISA"). A reorganized semantic network indicated the hierarchical relations among the concepts can be automatically created as shown in Fig.4d.

In another example, four concepts damage, overstress, shear overstress and inadequate in shear strength are related as shown in a semantic network of Fig.4e. After knowledge reorganization, a chain structure can be developed as shown in Fig.4f.

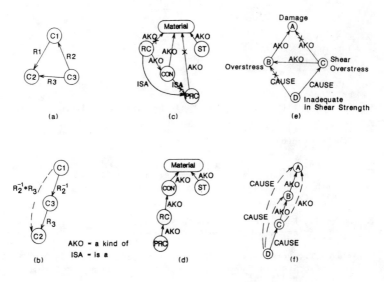

Figure 4 Triangle Model for Knowledge Reorganization

6. CONCLUSION

The development of the SPERIL-3, a knowledge-based expert system for structural damage assessment is introduced herein. The motivations for the development of this system are presented and discussed. Knowledge organization, knowledge representation, knowledge acquisition and machine learning in the SPERIL-3 are also described and discussed. A knowledge acquisition model called conditional knowledge is introduced and applied, which is further developed for the automation of knowledge acquisition. A triangle model is also developed for knowledge reorganization in the automation of knowledge acquisition.

ACKNOWLEDGMENT

This investigation is supported in part by the National Science Foundation through Grant No. CEE 8412569.

REFERENCES

1. Brown,C.B., and Yao,J.T.P., (1983) "Fuzzy Sets in Structural Engineering", Journal of the Structural Division, ASCE, Vol.109, No.5, pp. 1211-1225.

2. Carbonell,J.G., Michalski,R.S., and Mitchell,T.M., (1983) "An Overview of Machine Learning", in: R.S.Michalski, J.G.Garbonell, and T.M.Mitchell (Eds.) Machine Learning, Tioga, Palo Alto, CA, pp.3-23.

3. Cheng,Y., (1986) Combination and Acquisition of Conditional Knowledge, Ph.D. Thesis, School of Electrical Engineering, Purdue University.

4. Cohen,P., Davis,A., Day,D., Greenbery,M., Kjeldsen,R., Lander, S., and Loiselle,C., (1985) "Representativeness and Uncertainty in Classification Systems," AI Magazine, Vol.6, No.3, pp.136-149.

5. Davis,R., (1977) "Interactive Transfer of Expertise Acquisition of New Inference Rules," IJCAI 5, pp.321-328.

6. Ishizuka,M., Fu,K.S., and Yao,J.T.P., (1981) "Inexact Inference for Rule-base Damage Assessment of Existing Structures," Proceedings, Seventh International Joint Conference on Artificial Intelligence, pp.837-842.

7. Laird,J., Rosenbloom,P., and Newell,A., (1986) Universal Subgoaling and Chunking, Kluwer Academic Publishers.

8. Nguyen,T., Perkins,W., Laffey,T., and Pecora,D., (1985) "Checking an Expert Systems Knowledge Base for Consistency and Completeness," IJCAI-85, pp375-381.

KNOWLEDGE ACQUISITION 15

9. Ogawa,H., Fu,K.S., and Yao,J.T.P., (1985) "SPERIL-II: An Expert System for Structural Damage Assessment of Existing Structures," Approximate Reasoning in Expert Systems, Edited by M. M. Gupta, A. Kandel, W. Bandler, and J. B. Kiszka, Elsevier Science Publishers, North-Holland, pp. 731-744.

10. Scott,S.E., Zhang,M., and Zhang,X.J., (1986) "Human Information Processing Models and Their Application To Expert System," Technical Report IE-656-86, School of Industrial Engineering, Purdue University.

11. Winston,P.H., (1984) Artificial Intelligence, Addison-Wesley Publishing Company, Inc.,.

12. WJE Report (1987), "Fundamental Concepts of Building Damage," Prepared under NSF Grant No. CEE-8414075, Wiss, Janney, Elstner Assoc., Inc., Emeryville, CA.

13. Yao,J.T.P., Bresler,B., and Hanson,J.M., (1984) "Condition Evaluation and Interpretation For Existing Concrete Buildings," presentation at ACI 348/437 Symposium on Evaluation of Existing Concrete Buildings, Phoenix, AZ.

14. Zhang,X.J., (1985) "Expert System for Damage Assessment of Existing Structures," in: Some Prototype Examples for Expert Systems, edited by K.S. Fu, School of Electrical Engineering, Purdue University, Vol. 3, Chap.14.

15. Zhang,X.J., and Yao,J.T.P., (1986a) "Methodologies for Safety Evaluation of Existing Structures -- Literature Review," Technical Report No.CE-STR-86-28, School of Civil Engineering, Purdue University.

16. Zhang,X.J., and Yao,J.T.P., (1986b) "The Development of SPERIL Expert Systems for Damage Assessment," Technical Report No.CE-STR-86-29, School of Civil Engineering, Purdue University.

17. Zhang,X.J., and Yao,J.T.P., (1987) "Application of the Alpha-Beta Procedure in Decision Information Process," Technical Report No.CE-STR-87-34, School of Civil Engineering, Purdue University.

Expert System RAISE-1

Ruijin Chen*   Xila Liu**

Abstract

This paper describes the construction of the factor relation graphs for the reliability assessment of reinforced concrete frames and the fuzzy reasoning for the assessment. An expert system based on the factor relation graphs and the reasoning method is presented.

The system can consider the interaction between the main frame and the roof or walls. It can be used to assess the structural reliability, to diagnose the causes of the failure, and to classify the damage states.

This system is written in LISP and can be performed on IBM-PC computers.

1. Introduction

Building structures are exposed to various types of natural hazards. Meanwhile, materials in the structures are subject to deterioration and repeated application of loads. At times, it may be difficult to continue the proper usage of certain structures through the design life. Therefore, one of the important problems in structural engineering is to assess the reliability of existing structures. Because of the complexity and the uncertainty of the existing structures, few experts can make such assessment based on their intuitions and experiences.

Reliability for existing structures, discussed in this paper, includes the safety, the functional integrity, and the durability.

RAISE-1(Reliability Assessment In Structure Engineering-1), the expert system to be presented in this paper, is designed to assess the reliability of the structure shown in Fig.1.1 which consists of three sub-structures : the three-hinge reinforced concrete (R/C) frame, the R/C roof, and the enclosing walls.

2. Factor relation graphs

The complex problem of the assessment of structural reliability is decomposed into simpler sub-problems, the assessments of the safety, the functional integrity, and the durability, which are further decomposed into even simpler sub-problems or factors (Yao, 1985). This procedure is repeated until each factor becomes a primary one, which

*Graduate Assistant, Dept. of Civil Engineering, Tsinghua University, Beijing, 100084, China.
**Professor, Dept. of Civil Engineering, Tsinghua University, Beijing, 100084, China.

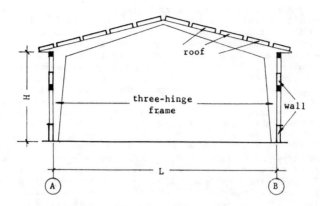

Fig.1.1 the R/C frame in RAISE-1

is obtained by the information collection directly.

A factor relation graph is proposed to represent the relations among factors created above (Lui and Chen, 1987.5). Fig.2.1 to Fig.2.5 show

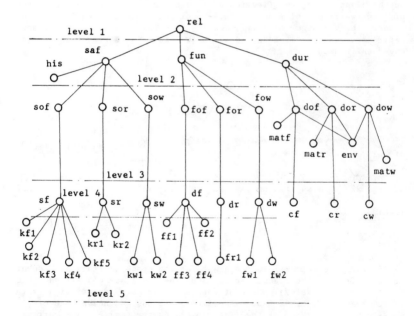

Fig.2.1 The factor relation graph for the reliability assessment

of R/C three-hinge frame(including roof system and surrounding walls). The reliability (level 1) of the structure is examined from its safety, functional integrity, and durability (level 2), which are the syntheses of the safety, functional integrity and durability of each sub-structure (level 3). They depend on the strengths, the deformations, cracking states, environment, and materials(level 4), where strengths, deformations, etc. can be checked or measured in detail(level 5), while the cracking states must be further decomposed. rel---reliability of the structure. saf---safety of the structure. fun---functional integrity of the structure. dur---durability of the structure. sof---safety of the frame. sor---safety of the roof. sow---safety of the walls. fof---functional integrity of the frame. for---functional integrity of the roof. fow---functional integrity of the walls. dof---durability of the frame. dor---durability of the roof. dow---durability of the walls. sf---strength of the frame. sr---strength of the roof. sw---strength of the walls. df---deflection of the frame. dr---deflection of the roof. dw---deflection of the walls. cf---cracking state of the frame. cr---cracking state of the roof. cw---cracking state of the walls. kf1---flexural safety factor of the beams. kf2---shear safety factor of the beams. kf3---safety factor of the columns. kf4---safety factor of the rigid joints. kf5---safety factor of the top-hinge. kr1---safety factor of the roof slabs. kr2---reliability of the connection between the roof and the frame. kw1---reinforced steel in the walls. kw2---rationality of the wall sizes. ff1---deflection of the beams. ff2---deflection of the columns. ff3---inclination of the frame in plane. ff4---inclination of the frame out of plane. fr1---deflection of the roof slabs. fw1---inclination of the walls out of plane. fw2---inclination of the walls in the plane. his---usage history. matf---material state of the frame. matr---material state of the roof. matw---material state of the walls. env---environment.

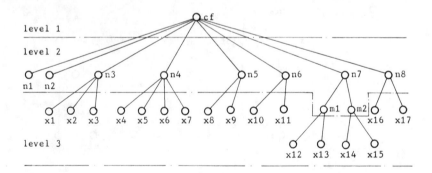

Fig.2.2 The factor relation graph for the cracking state assessment of the frame. The cracks (level 1) on the frame are classified into eight types (level 2). Some of them need a minute description with their breadths, lengths, and sometimes positions and numbers(level 3). n1---vertical cracks near the rigid joints on the beam. n2--- vertical cracks on the columns. n3---cracks in the shear region. x1--- breadth

of n3.   x2---position of n3.   x3---number of n3.   n4---cracks in the flexural region. x4---breadth of n4. x5---length of n4. x6--- number of n4.   x7---position of n4.   n5---cracks due to concrete shrinkage. x8---breadth of n5. x9---length of n5. n6---cracks due to steel rust. x10---breadth of n6.   x11---length of n6.   n7---cracks at the rigid joints.   m1---type 1 of n7.   m2---type 2 of n7.   x12---breadth of m1. x13---length of m1.   x14---breadth of m2.   x15---length of m2. n8--- cracks at the top hinge joint. x16---breadth of n8. x17---length of n8.

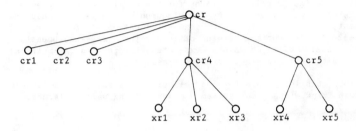

Fig.2.3 The factor relation graph for the cracking state assessment of the roof. Cracks on the roof may appear in 5 type. Some of them are determined by their breadths. The other are described with not only their breadths but also lengths or even numbers.   cr1---cracks between roof slabs and walls. cr3---cracks at the end of the slabs.   cr5--- cracks due to concrete shrinkage. xr2----length of cr4.   xr4--- breadth of cr5. cr2---cracks between slabs. cr4---cracks on the surface of slabs. xr1---number of the cr4.   xr3---breadth of cr4. xr5---area of cr5.

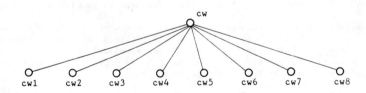

Fig.2.4 The factor relation graph for the cracking state assessment of the walls. There may be eight types of cracks on the walls resulting from different causes. cw1---crisscross cracks.   wc2---cracks well-distributed horizontally and vertically.   wc3---"/ \" form cracks symmetric to the columns of the frame. wc4---"\ /" form cracks symmetric to the columns of the frame. wc5---"//" or"\\" single diagonal cracks. wc6---double diagonal cracks. wc7---contract cracks. wc8---other type cracks.

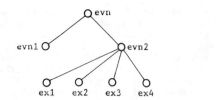 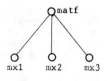

Fig.2.5 The factor relation graphs for the environment assessment and the material assessment. env---environment. env2---non-chloride environment. ex2---humidity. ex4---evaluation of corrosive substances. mx1---surface state of concrete. mx3---carbonation. env1---chloride. ex1---PH value. ex3---temperature. matf---material state of the frame. mx2---corrosion of steel.

the factor relation graphs for the reliability assessment of a R/C three-hinge frame with a roof system and surrounding walls attached.

3. Some definitions for reasoning

Definition 3.1: When one node is affected directly by several other nodes underneath, it is called the parent with respect to the nodes below. The nodes underneath are called children with respect to the node above. The top node, the one with no parent, is called the goal. The bottom nodes, those with no children, are called the primary nodes. Furthermore, those with not only parent but also children, are called intermediate nodes.

Let x be a node and Fx be the factor denoted by x.

Definition 3.2: Let $LX=[0,1]$ and lx ($\in LX$) correspond to an integral state of Fx. For example, 0 and 1 correspond to 'limit state' and 'perfect state', respectively. Then LX is called the assessment universe of x, and lx the assessment value of Fx.

Definition 3.3: (1) When node x is a intermediate node or a goal node, its domain, denoted as X, is defined as the Cartesian production of its children' relevant assessment universes. (2) When node x is a primary node, its domain, X, is defined as a range that includes all possible input values for the node.

Definition 3.4: $\underset{\sim}{A}X$ is called a fuzzy subset of node x, iff its membership function is the mapping

$$X \longrightarrow LX \qquad (3.1)$$
$$x \longmapsto \mu_{\underset{\sim}{A}X}(x)$$

where x X.

Thus any node, x, for example, in a factor relation graph, can be assigned with a six element set

$$(Fx,\ x,\ X,\ LX,\ AX,\ \mu_{\underset{\sim}{A}X}(x)) \qquad (3.2)$$

to describe the information of the node,
where Fx = the factor denoted by node x
       x = not only the symbol of the node but also the variable
           value in X
       X = domain of node x
       LX = assessment universe of node x
       $\underset{\sim}{A}X$ = a fuzzy subset of node x
       $\mu_{\underset{\sim}{A}X}(x)$ = membership function of $\underset{\sim}{A}X$.

4. About the reasoning method

The reasoning method used in RAISE-1 is called a membership function method (Chen and Lui, 1987.3), In which a continuous membership function is used to reflect the transition of assessment values from 0 to 1. A variable weighting (Wang and Zhang, 1985) is employed to interpret the complex relations among the children and between the children and the parent.

Let x1, ..., xk be the children of y (Fig.4.1). The reasoning in RAISE-1 is to deduce ly from lx1, ..., lxk, Where lxi ∈ LXi, i=1, ..., k, and ly ∈ LY, and LXi and LY are the assessment universes of xi and y , respectively. This process can be represented as

Fig.4.1 x1, ..., xk are the children of y

$$ly = \mu_{\underset{\sim}{A}Y}(y) \qquad (4.1)$$

where y = (lx1, ..., lxk) and $\mu_{\underset{\sim}{A}Y}(y)$, the membership function as defined in definition 3.4 , is a function of lx1, ...,lxk.

In order to construct $\mu_{\underset{\sim}{A}Y}(y)$, y's children are divided into characteristic children and modificatory children. The classification is based on the following two properties of characteristic children.

Property 4.1: Within y's domain, Y, (not including its boundary), for any characteristic child xi, If lxi ⟶ 0, then ly ⟶ 0 no matter what values lx1, ..., lxk take within (0,1).

The essential part of the parent's assessment value, denoted by ly, according to experts, can be obtained from characteristic children first. This process is called the first step reasoning.

Property 4.2: Assume that y has k children, in which x1, ..., xp are characteristic children and xp_{+1} , ..., xk are modificatory children. Then

$$\min(lx1, ..., lxp) \leqslant ly \leqslant \max(lx1, ..., lxp) \qquad (4.2)$$

After the classification of children, comes the fist step reasoning, which can be conducted as follows:

$$l_y = \sum_{i=1}^{p} \lambda_i \, lx_i \qquad (4.3)$$

$$\sum_{i=1}^{p} \lambda_i = 1 \qquad (4.4)$$

where λ_i is the function of $lx1, \ldots, lxp$, and is called the variable weighting or weighting function.

On the basis of property 4.2, when $lxi=0$, the boundary condition for λ_i is

$$\lambda_i = 1 \qquad i = 1, \ldots, p \qquad (4.5)$$

When $lx1 = \ldots = lxp$, λ_i becomes the ordinary weighting:

$$\lambda_i = \lambda_i^0 \qquad i = 1, \ldots, p \qquad (4.6)$$

λ_i^0 is given by experts.

The solutions satisfying Eqs.(4.4), (4.5) and (4.6) may be infinite. In practice, other conditions such as the regularity of weighting functions varying with their independence variables should be taken into account also.

The modificatory reasoning can be found in detail in Ref.(Chen and Lui, 1987.3).

5. About RAISE-1

5.1 Structure Of RAISE-1

RAISE-1 consists of ten subsystems (called trays) : TYAY0 --- TRAY9 (Fig.5.1).

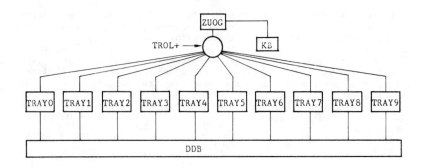

Fig.5.1 the structure of RAISE-1

ZUOG, with a general knowledgebase KB which can be used by any subsystem, introduces RAISE-1 itself to the user. TRAY0 and TRAY1 are the smallest subsystems. TRAY0 inquires about the condition and the history of the structure and gives an estimation. TRAY1 evaluates the current material state of the structure. TRAY2 examines the environment condition. TRAY3 assesses the durability of the three-hinge frame. TRAY4 assesses the safety and the functional integrity of the three-hinge frame. TRAY5 examines previous primary data and intermediate conclusions, and makes second questions about the three-hinge frame, if necessary. TRAY6 diagnoses causes of the failure of the frame. TRAY7 deals with the roof system. TRAY8 deals with enclosing walls. TRAY9 assesses the reliability of the whole structure including the three-hinge frame, the roof and wall systems. In TRAY9 some recommendations and the damage classification are also given.

TROL+ provides the interfaces among subsystems, allowing the user to move from one tray to another without exiting from RAISE-1. All primary data and intermediate conclusions are stored in the dynamic database DDB. DDB works as a common database for all trays.

5.2 A tray

Take TRAY3 as an example to show the structure of a tray. It consists of four components: an interpretation system, an inference machine, a partial knowledgebase KB3, and a displaying block BLOCK3 (Fig.5.2).

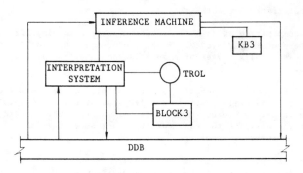

Fig.5.2 the structure of TRAY3

The interpretation system, consisting of a set of 'ASK' functions, is the interface between the user and RAISE-1. It can be used to ask for information by the inference machine or to modify primary data by the user. For example, when inference machine searches out x1 (Fig.2.3), it calls function ASK1 as follows :

The widest breadth of the cracks in the shear region
x1=? mm
R>

Where 'R>' is the prompt in RAISE-1. The partial knowledgebase KB3 consists of a set of membership functions or other defined LISP functions arranged in some manners. The inference machine is a procedure, which controls the interpretation system and knowledgebase. First, in the factor relation graph it searches from the top to the bottom using a strategy similar to generation-test one (Hayes-Roth etc., 1983), pruning branches on the factor relation graph so as to create a 'real factor relation graph' for a certain structure. For example, when it searches out n4 (Fig.3.2). it calls function ASK5 to ask the user:

Are there any cracks in flexural region ? (Y/N)
R>

If it receives a negative answer, it gives up asking anything about x4, x5, x6, x7, and turns to n5. That is, it prunes the branch including n4,x4, ...,x7. Then it infers in the 'real factor relation graph ' from the bottom to the top. BLOCK3, a displaying block consisting several slides, provides the user with an opportunity to refer to the primary data, intermediate conclusions, and the reasoning relations between factors. It also provides the user an opportunity to modify the primary data. BLOCK3 is controlled by TROL, which provides the user some keychord commands to use BLOCK3.

5.3 Characteristics of RAISE-1

The decomposition of RAISE-1 into ten subsystems simplifies each of the interpretation system, the inference machine, and the knowledgebase. 512k bytes of memory is enough to operate it on IBM-PC computers.

In RAISE-1, knowledgebase is divided into a general knowledgebase(KB) and several partial knowledgebases(KBi, i=1,...9). Most knowledge in KBi is stored in the external memory. When needed, it is read from the diskette to reason or to answer some questions. It does not take the internal memory.

RAISE-1 keeps testing the input data. Whenever it catches any invalid data, it would remind the user of the data type to be expected and require him/her to try once again. For example,

The number of the cracks in the flexural region ?
R> a
Wrong input ! An integer is expected. Please try again.
R>

So RAISE-1 is very easy to handle. It is not necessary to require the user to be a programmer.

In RAISE-1 there is no such LISP function as 'WHY' or 'HOW', which has been used in many other expert systems such as in MYCIN for user to ask the reasoning relations among factors, however the displaying block BLOCKi (i =2, ..., 9)not only has the same effect but also provides the user an opportunity to modify the primary data without exiting from RAISE-1, which helps the user to determine the strengthening plan. When the RAISE-1 works out a conclusion that the structure or its elements are failure, the user may decide a plan to improve the structural reliability (Fig.5.3).

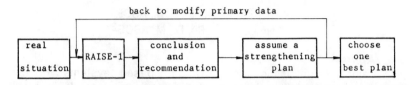

Fig.5.3 making a strengthening plan in RAISE-1

The software of RAISE-1 can be provided in the Civil Engineering Department of Tsinghua University (Beijing).

6. Reference

[1] Chen, R.J., and Liu, X.L., "a Method of Fuzzy Reasoning Representation in the Reliability Assessment of R/C Structures", No. CE-ESS-87-01, Dept. of Civil Engineering Tsinghua University, 1987.3. (in Chinese).
[2] Hayes-Roth, F., Waterman, D., and Lenat., D., eds, "Building Expert systems", Addison-Eesley, 1983.
[3] Liu, X.L. , Chen, R.J., "the Reliability Assessment of R/C Structures and its Factor Relation Graphs", No. CE-ESS-87-03 , Dept. of Civil Engineering, Tsinghua University,1987.5. (in Chinese).
[4] Wang, P.Z., Zhang, D.Z., "Fuzzy Decision----Theory and Application", Beijing Normal University, 1985.6. (in Chinese).
[5] Yao, J.T.P., "Safety and Reliability of Existing Structures", Pitman Advanced Publishing Program, 1985.

Using PROLOG on a Macintosh to Build an Engineering Expert System

W. M. Kim Roddis, P. E., M. ASCE[1]
Jerome Connor, Ph. D., M. ASCE[2]

1. Abstract

A major design issue for knowledge based systems for civil engineering is deliverability to practicing engineers. It is highly desirable for a fieldable version to run on micro-computer hardware of a type with wide usefulness in an engineering office. For ease of use and high rate of information communication, a multi-window interface that supports alphanumerics, menus, and high quality black and white graphics is desirable. In addition to these hardware design criteria, the software required should be flexible and cost effective. An example of an engineering expert system that was developed to meet these deliverability specifications is CRACK, a knowledge based system for the problem domain of fatigue and fracture in bridges. The primary objective of this paper is to explain how the relatively inexpensive hardware and software tools of the Macintosh personal computer line and the PROLOG language can be used to solve a significant engineering problem using a knowledge based approach.

2. Context

Over the past five years the Civil Engineering Department at MIT has become increasingly involved in the application of Artificial Intelligence techniques to engineering problem solving, especially in the area of knowledge based systems. Projects have addressed diverse problem areas, called problem domains. These domains have included generative tasks, such as design of steel members and type studies for bridge preliminary design, as well as diagnostic tasks, such as condition assessment of concrete bridge decks. The particular project which motivated this paper deals with fatigue and fracture in steel plate girders. In the course of developing this extensive prototype engineering expert system, two tools were

---

[1] Fannie and John Hertz Foundation Fellow, Civil Engineering Department, MIT, Cambridge, MA, 02139.
[2] Professor of Civil Engineering, MIT, Cambridge, MA, 02139.

used ìich have wide applicability for civil engineering. The
softwaı tool is the PROLOG programming language and the
hardware tool is the Macintosh line of personal computers.

This paper is organized in the following manner. First, an argument is made for the advantages of a knowledge based approach. The terminology of knowledge based systems, expert systems, shells is clarified. A system architecture to address the problem domain of cracking in bridges is presented. After the required software and hardware needs are established, the selected software tool, PROLOG, is introduced. The chosen hardware platform, the Macintosh family, is shown in the light of general engineering usefulness. Finally, the major points with broad applicability to similar projects are summarized.

3. Knowledge based systems, expert systems, and shells

The terms knowledge based system and expert system are not precisely defined or consistently used. This paper uses the term knowledge based system to apply to programs with explicit knowledge representation and separation of the knowledge base from the inference mechanism. The knowledge base contains the domain information made up of facts, rules describing relations or physical laws, and methods for problem solving in the domain. The inference engine uses domain independent methods of drawing conclusions by manipulating and using the knowledge base. The appeal of a knowledge based approach is the explicit statement of the knowledge used to solve a problem. This allows multiple uses of the same knowledge and incorporation of disparate kinds of knowledge (numeric/symbolic, rules of thumb, algorithms).

Here, the term knowledge based system is not used as an equivalent to the term expert system. An expert system is a subset of knowledge based systems with the added capability of explaining its behavior [Bratko]. In addition, the term expert system implies a certain level of performance in the delivery system, but for the context of this paper the term expert system is meant to emphasize the fact that a knowledge based system has an explanation/justification capability. Explanation is a crucial feature of an expert system, used to validate the system's reasoning during a consultation. An expert system should be able to explain its behavior and justify its requests for information and its problem solutions. As a minimum it should be able to explain **why** it asks particular questions and **how** it reached its conclusions.

Traditio   numeric computing techniques rely on large amounts of non-algorithmic knowledge for correct program use. This knowledge is either represented implicitly in the source code, or not represented in the program at all, except possibly as passive comments. Implicit representation isn't enough. Numeric computation should be produced in a way that explicitly represents algorithmic and heuristic knowledge about the programs. This explicit knowledge should be manipulated to make the "black boxes" of number crunching transparent to the user. Explicit representation is required for automatic explanation of the system's behavior and justification of conclusions. It is this explanation/justification feature that can be exploited to make expert systems **reliable**, an area of great importance to the practicing engineer.

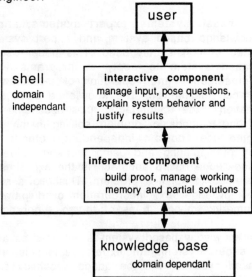

Figure 1: Simple Shell Structure

Expert systems have a domain dependant part, the knowledge base and a domain independent part, called the expert system shell. The shell consists of two main parts, the interactive component and the inference component as shown in **Figure 1**, the interactive part handles all the user interface tasks for adequate communication between the system and the user. The inference part contains the actual inference engine for reasoning, the context of the particular consultation, the incomplete solutions as the problem proceeds, and the connections to the knowledge base. Although shells are in principle domain independent, in practice applications must be closely related to satisfactorily use the same shell.

4. Architecture of CRACK

CRACK (Consultant Reasoning About Cracking Knowledge) [Roddis] is a knowledge based system for the problem domain of fatigue and fracture in bridges. The system architecture is capable of supporting engineering reasoning about both the overall problem domain at a relatively high level of abstraction and the narrow subdomains at a deeper level of knowledge. The system links heuristic and quantitative simulation levels into an integrated framework by using a middle layer of approximate models. The ability to direct search by reasoning with a simplified engineering model and then verifying, revising and refining the rough model by recourse to more exact analysis lies at the heart of engineering problem solving. Application of this general methodology to the specific domain of fatigue and fracture in steel bridges is used to prove the validity of the concept and to develop and demonstrate the power of this approach. The approach is applicable to a wide variety of engineering problems.

There are many reasons why this problem domain is well suited for use of a knowledge based approach, but three deserve special mention. First, this is a practical problem where the knowledge exists but is frequently not utilized. Second, the knowledge is of diverse types (statistical, heuristic, engineering principles, etc.) but circumscribed and well contained so that it is possible to do a complete coverage of the knowledge needed to solve the problem. Third, there are multiple uses of the same knowledge; for analysis of failures, for determine causes of distress and fixes, for predicting remaining service life, and for verification and optimization of design.

First generation knowledge based systems are primarily rule based. This technique works well for domains predominated by compiled experience expressible in empirical associations such as: **Previously, when A and B held, C was also found to be true.** Rules alone do not work well for domains relying heavily on understanding of structure, function, and causality typified by knowledge like: **If A and B are true, C follows because of the way this device works** [Davis].

Engineering domains frequently are of this character, making rule-based reasoning alone an inadequate tool. For this reason, the three layered structure shown in **Figure 2** was developed for CRACK. The three layers unite the disparate approaches of rule-based reasoning, qualitative simulation, and numerical analysis. The top layer uses a backward chaining rule-based paradigm to process domain knowledge that is predominantly declarative. The mid-layer uses qualitative causal structure descriptions to express the relationships among the physical parameters. The root layer is made up of quantitative simulators to represent domain knowledge that is largely procedural.

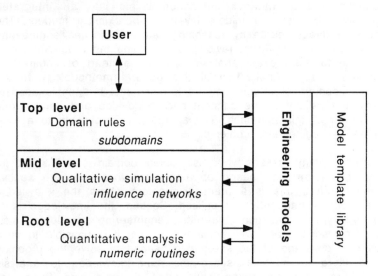

Figure 2: Layered Structure of CRACK

Each method is a better tool for a different phase of the engineering problem solving process. Rules are good for generating hypotheses and defining initial areas in which to search for a solution. Qualitative reasoning is suited to enumerating possible behaviors and focusing in on the most promising engineering models. Quantitative methods are best to resolve ambiguous behavior and arrive at precise answers. The power of the entire computer application is increased by using the right tool for the task at hand.

5. System Requirements

The requirements analysis stage of a project determines if the planned project will satisfy a real need while being technically and economically feasible. As with any other programming project, the development of an engineering expert system involves the comparison and selection of an appropriate set of software and hardware tools. To be able to select these tools, the desired characteristics of the assembled system must be defined. For our project, the key goal having the widest effect on system selection was deliverability to practicing engineers. A central motivation for our project was to demonstrate how to realize the promise of advanced computing techniques from Artificial Intelligence in the context of day-to-day work in even a small engineering office. Our list of desired system characteristics is therefore headed by deliverability. The development system does not need to be the same as the target delivery system, but the development process must result in a product able to be ported to the final field delivery system.

The core list of system requirements is as follows. The fieldable version must run on relatively inexpensive micro-computer hardware of a type with wide usefulness in an engineering office. The system must be user friendly: easy to learn, convenient to use, with a consistent operating interface. A multi-window interface that supports alphanumerics, menus, and high quality black and white graphics is needed to obtain a high rate of information communication. The software required should be flexible and cost effective. If a high level language package must be purchased, its usefulness is greatly enhanced if it supports interlanguage communication to allow connection to other engineering software.

The conventional wisdom when undertaking purchase of a computer system is to first find the software that is appropriate for applying a computer to the work and only then looking into the hardware. However, a typical engineering office can ill afford to purchase a computer system dedicated only to knowledge based systems. For the engineer, the hardware should ideally serve as a satisfactory platform for software. that addresses the specific area of knowledge based applications while providing general computing capability. This not only allows the hardware to be used for other tasks in the office, a practical necessity, but also makes it possible to take advantage of rapid advances in software and to maintain flexibility for addressing new problems not foreseen in the original definition of needs.

The new generation of personal computers which have become available within the last few years have made it possible to exploit some of the advanced computer methods developed for Artificial Intelligence. To make tools like knowledge based systems viable for widespread use in civil engineering this transfer from an expensive special purpose hardware platform like a large LISP machine workstation to microcomputers was an essential step.

This paper makes no pretense of undertaking an exhaustive survey and evaluation of personal computer based Artificial Intelligence tools. Given the current speed of development and relative immaturity of the field, the short lead time for specialty magazines makes them the proper format for such reviews [Citrenbaum] [Freedman]. The lessons here are pragmatic ones, based on considered in-depth application to one problem domain, carefully selected to be representative of a large, crucial, and central class of engineering applications.

To choose a software tool, available commercial and research shells were first investigated. In general, the multiplicity of required features and possible hardware environments make it impossible for one single shell to satisfy all possible needs without becoming unwieldy. After experimenting with in-house software tools and several commercial products, the conclusion was reached that the shells available in mid 1986 when the project's implementation began did not provide a convenient method of integrating the disparate reasoning mechanisms of rules, causal models, and numeric analysis which is essential for engineering problem solving.

For this reason, a shell which did have the desired capabilities was built. There are great advantages to using an in-house shell, especially for a research project. The size of the shell is kept small since only necessary features are included. Changes can easily be made as the project development proceeds. The shell for CRACK is built on top of the intrinsic theorem prover in PROLOG which is already a powerful deductive mechanism. Many examples can be found for using PROLOG for shells [Hammond], [Sterling], [Bratko]. A bare PROLOG program is not a knowledge based system, but the language is a good tool for implementing a shell. A shell is required for the separation of knowledge from the algorithms that use the knowledge required of a knowledge based system. A shell is also necessary to produce the explanation capability required of an expert system. The native PROLOG inference engine is not sufficient to produce problem solving behavior which appears sensible to the user without the addition of a shell that specifies the desired problem solving procedures.

If construction of an in-house shell is not feasible, then a commercial shell must be chosen for knowledge based applications. The first step in selecting a shell is a needs analysis. This is mainly a definition process, deciding what features the shell must have to address the specific domain. The next step consists of gathering information from the literature, including magazines for up-to-date product reviews, and from experienced people and vendors. This identifies interesting software which must be tried out. One example of a place to start is the AIExpert magazine publication intended to provide introductory overview articles with pointers into the literature for pursuing an area in more depth [Citrenbaum] [Freedman].

6. PROLOG

The programming language PROLOG has may desirable features. Its goal oriented style of programming is generally clear, transparent, concise, and modifiable. There are two basic programming styles. The procedural or imperative approach describes the behavior needed to achieve the desired result. The declarative or descriptive approach gives a descriptive definition of a set of relations or functions to be computed. This procedural/declarative dichotomy can be clarified by making the analogy to material specifications (procedural, describe how to do task by supplying specific items) versus performance specifications (declarative, describe what requirements the end product must meet) in engineering. PROLOG facilitates a declarative programming style. The emphasis is on what needs to be done by the program, not on how to do it. This allows concentration on the knowledge rather that the algorithms.

PROLOG (programming in logic) was an academic mathematical and research tool developed in Europe [Clark84]. Robert Kowalski at Edinburgh contributed crucial theoretical work in the early 1970's. The computer language was first implemented by Alain Colmerauer at Marseilles in 1972. The major step of producing an efficient interpreter/compiler was taken by David Warren at Edinburgh in 1977. This implementation established the Edinburgh syntax (also known as the DEC-10 syntax) which has become the industry standard as the most wide-spread and hence most compatible and portable of the various computer languages in the PROLOG family. Clearly, compatibility and portability are highly desired, since application growth and change in available host machines must be anticipated. The visibility of PROLOG was greatly magnified by the public attention given to the 1981 initiation of the Japanese Fifth Generation Computer Project, which gives PROLOG a central role.

For a brief introduction to the deductive formalism on which PROLOG is based see [Kowalski]. **Figure 3** illustrates the central idea of inference in PROLOG. A program is made up of assertions and implications expressed in symbolic logic. Inference is done using resolution (a generalization of modus ponens).

Modus ponens: $(X \land (X \rightarrow Y)) \rightarrow Y$

Which is to say: Given the assertion **X** and the implication **X implies Y**, the assertion **Y** can be inferred

Example: Socrates is an ancient Greek.
If someone is an ancient Greek, then that person is dead
Therefore, Socrates is dead.

PROLOG: given
ancient_greek(socrates).
dead(X):- ancient_greek(X).
infer dead(socrates).

Figure 3: Inference in PROLOG

What is of immediate interest to most engineer/programmers is not the theoretic underpinnings, but how the language is used. One text, used by many to learn PROLOG [Clocksin], was largely alone in the field for several years. Recently, several more extensive and up-to-date introductory texts have appeared [Sterling] [Bratko].

It was pointed out that the implementation of an efficient compiler was a major step. Those familiar with only traditional programming languages such as FORTRAN and C may not be aware of the difference between an interpreter and a compiler. An interpreter reads a source program and translates it into machine language as it executes it, step by step. Interpreters allow stopping the program at any point, making changes, and continuing the same execution. This flexibility has the price of being time and space intensive. A compiler translates the whole source program to machine code, creating an object code. The object code is then separately executed to make the program run, resulting in much greater efficiency but making changes harder since the source must be edited and recompiled. The space and time requirements of real (non-academic or purely research) applications require the efficient execution of compiled PROLOG.

Pattern directed rule based programming is very natural in PROLOG. Built in features of the language (pattern matching, automatic backtracking, and symbolic computation) combined with its well understood deductive formalism make it a good choice for development of an expert system. PROLOG, as a full high level language, does have imperative features (assert, delete, cut, file input/output, etc.), so a custom tailored programming environment can be built up. Extensive numeric processing is not convenient or efficient in PROLOG, so for engineering applications foreign functions calls are a valuable extension. One of the attractive features of PROLOG is its availability on small machines, allowing the development and delivery of sophisticated knowledge based systems without large, costly, dedicated hardware.

7. Apple Macintosh personal computers

Having selected the software tool, PROLOG, the hardware configuration must be determined. Apple Computer's Macintosh line of personal computers is not new. The familiar configuration consisted of a 128K Mac, 512K Mac, or 512K Extended Mac, usually accompanied by an external disk drive, printer and frequently a modem. With the introduction first of the 1 megabyte Macintosh Plus and then the expandable Macintosh SE and Macintosh II, the Mac line became a promising hardware platform for technical work like engineering, delivering high value per dollar. The Mac line's user friendly visual interface, strong graphics capabilities, and connectability into networks supporting multiple kinds of computers, not just Apple products, are all attractive features. The speed and customizability of the Macintosh II, whose standard features include a math coprocessor and six expansion slots, make it the most attractive choice for engineering CAD/CAM/CAE and architectural applications [The CAD/CAM Journal]. The use of Macintoshes in the engineering community is rapidly expanding, leading to a positive feedback of more users supporting an expanded software market, which in turn encourages more users.

The availability of high quality engineering software if thus growing quickly.

Although specialized magazine publications [MacUser] should be referred to for more timely information, it is worth presenting a summary of the hardware found useful for application of artificial intelligence techniques to engineering problem solving in developing CRACK: the Macintosh Plus, Macintosh SE, and Macintosh II products The space requirements on the 512K Mac was found to be inadequate after the initial stages of the project. The practical minimum for developing these types of applications appears to be 1 megabyte. For delivery, this limit could perhaps be pushed to 512K if the developed expert system was tailored for a small machine by removing the development environment features.

The Macintosh Plus, costing under $2000, is the least expensive and least powerful, with 1 megabyte of memory, 800K internal floppy drive, hard disk connector, and a closed architecture making expansion relatively inconvenient. The Macintosh SE , in the $3000 range, and the Macintosh II, in the $4000 to more than $5000 range depending on options selected, both come with 1 megabyte of memory expandable to 2, 4, or 8 megabytes (the Mac II can also be extended beyond all reasonable bounds to 1.5 gigabytes using expansion slots). Hard disks, in the 20 to 80 megabyte range, either internal or external are usually selected. The Macintosh SE has a 9" built-in monochrome monitor. The Macintosh II has a separate 12" monitor, either monochrome or color.

A major objection to the Macintosh computers has been the small screen size. We have found the high resolution small screen to be surprisingly acceptable, although the 12" Macintosh II color monitor is more convenient and is obviously more appropriate for demonstrations to another person since more windows can be legibly shown. Software screen extenders overcome a lot of the confinement of the small screens without the cost of a large screen monitor. The inherent speed of the Macintosh II with its math coprocessor is of great use for heavy duty number crunching, although the Macintosh SE and Macintosh Plus can have expansion boards added to increase their calculation speed, for an increased cost.

8. Conclusions
A central contribution of the system development for CRACK is the multilevel approach. The strategy of using an intermediate

qualitative simulation layer manipulating first order engineering models to connect a predominantly heuristic and symbolic rule-based top layer with a largely procedural and numeric quantitative root layer is applicable to a wide variety of engineering problems. The system represents the development in a particular domain of a flexible and robust engineering problem solving tool capable of describing a multifaceted engineering problem solution and of explaining the solution process.

The key criterium of deliverability to the practicing engineer led to the choice of the PROLOG language on Macintoshes. The particular software tool selected was LPA MacPROLOG [Clark88]. This product provides a unified environment with the complete power of PROLOG, the user friendly Macintosh philosophy interface (mice, menus, icons) with multiple windows, and high resolution graphics as shown schematically in **Figure 4**. An interface is provided to both C and Pascal, with the ability to call a compiled program. Future planned enhancements include color, database interface tools, and one of particular interest in the context of this paper, an expert system toolkit.

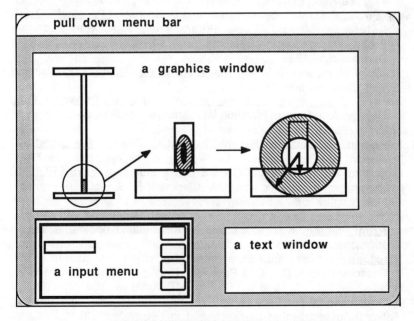

Figure 4: CRACK interface

Running full featured PROLOG on a powerful personal computer is a method of turning the potential of knowledge based and expert systems into solid solutions to engineering problems. The relatively inexpensive hardware and software tools of the Macintosh personal computer line and the PROLOG language can be used to solve a significant engineering problem with a knowledge based approach.

9. Acknowledgements

This work was supported by the Fannie and John Hertz Fellowship Foundation.

10. References

Bratko, Ivan, PROLOG Programming for Artificial Intelligence Addison-Wesley, Reading, Massachusetts, 1986.

Citrenbaum, R., J. R. Geissnam, and R. Schultz, "Selecting a Shell", A I Expert 500 Howard St., San Francisco, CA, Vol. 2 No. 9, September, 1987.

Clark, K. L, and F. G. McCabe, micro-PROLOG: Programming in Logic, Prentice-Hall, Englewood Cliffs, New Jersey, 1984.

Clark, K. L, F. G. McCabe, N. Johns, and C. Spenser, LPA MacPROLOG Reference Manual, Logic Programming Associates, London, Programming Logic Systems, 31 Crescent Drive, Milford, Connecticut, 1988.

Clocksin, W. F., and C. S. Mellish, Programming in PROLOG, Springer-Verlag, New York, 1981.

Davis, R., Expert Systems: Where Are We? And Where Do We Go From Here?, A. I Memo No. 665, MIT Artificial Intelligence Laboratory, June 1982.

Freedman, R.,"Software Review: Evaluating Shells", A I Expert 500 Howard St., San Francisco, CA, Vol. 2 No. 9, September, 1987.

Hammond, P., "micro-PROLOG for Expert Systems" in micro-PROLOG: Programming in Logic , by K. L. Clark and F. G. McCabe, Prentice-Hall, Englewood Cliffs, New Jersey, 1984.

Kolwalski, R., "Logic as a Computer Language" in Logic Programming, K. L. Clark and S. A. Tarnlund editors, Academic Press, London, 1982.

MacUser, "The Complete MacUser Systems Guide", Ziff Communications Co., One Park Ave, New York, September, 1987.

Roddis, W. M. K., and J. Connor, "Qualitative and Quantitative Reasoning about Fatigue and Fracture in Bridges" in Coupling Symbolic and Numerical Computing in Knowledge Based Systems,

J. S. Kowalik and C. T. Kitzmiller editors, North-Holland, Amsterdam, 1988.

Sterling, L., and E. Shapiro, <u>The Art of PROLOG</u>, MIT Press, Cambridge, Massachusetts, 1986.

<u>The CAD/CAM Journal: For The Macintosh Professional</u>, Koncepts Graphic Images Inc, 16 Beaver St., New York, Vol. 1 No. 5, October/November,1987.

SDL: AN ENVIRONMENT FOR BUILDING INTEGRATED STRUCTURAL DESIGN EXPERT SYSTEMS

Y. Paek and H. Adeli, Member, ASCE[*]

Abstract

A Structural Design Language (SDL) has been developed in INTERLISP environment for building coupled knowledge-based expert systems for integrated structural design problems. The complex body of knowledge needed for detailed design of a structure is fractionated into smaller and manageable knowledge sources which are organized into a hierarchy of cooperating conceptual specialists. SDL has been used to develop an expert system for integrated design of steel building structures consisting of moment-resisting frames.

Introduction

The process of detailed design of a structure made of a large number of components is quite involved. Intuition, judgment, and previous experience has to be used for selecting the right values for the design parameters. Further, considerable amount of numerical processing is involved in the detailed design of a structure in addition to heuristics. Therefore, an expert system for detailed design of a class of structures should be a coupled system in which AI symbolic processing and conventional numerical processing are integrated. Research in the area of coupled expert systems is relatively new (Kitzmiller and Kowalik, 1987, Adeli and Al-Rijleh, 1987, Adeli and Balasubramanyam, 1988a&b). AI languages such as Prolog and LISP have been developed primarily for non-numeric symbolic processing.

Among various dialects of LISP, INTERLISP seems to be the most highly developed implementation. INTERLISP has the rich characteristics of Conversational LISP (CLISP) interpreter, powerful error correction facility, and extensive library (Adeli and Paek, 1986a&b). INTERLISP has a "spaghetti" stack in which program contexts are stored (Rich, 1983). The spaghetti stack is in fact a tree structure where parallel contexts may be stored in such a way that control can be trasferred back and forth among them. Another feature of INTERLISP is the DWIM (Do What I Mean) facility that can be used for debugging the program. It can correct the misspelled words, for example.

INTERLISP does not have all the built-in procedures necessary for numerical processing, such as two dimensional arrays. But, it provides an environment for creating specialized languages. Therefore, to facilitate the development of expert systems for integrated design of structures we have developed a Structural Design Language (SDL) in

---

[*]Department of Civil Engineering, The Ohio State University, Columbus, Ohio 43210.

INTERLISP environment. The integrated design includes the preliminary design, structural analysis, design of members, design of connections, and computer-aided drafting of the final design. Our broader motivation is to develop an expert system shell for structural design problems. It should be noted that a large number of commercial expert system shells have been developed in recent years (Adeli, 1988, Moselhi and Nicholas, 1988). But, practically all of them are most suitable for solution of diagnosis problems.

Knowledge Representation

The knowledge necessary for structural design is classified into three categories: static knowledge, dynamic knowledge, and graphical knowledge (Paek and Adeli, 1988).
Static knowledge is defined as the knowledge necessary for representing the physical structure, its components, and their topology. As an example, for a structural frame in a building the static knowledge includes the knowledge of geometry and properties of beams and columns and the knowledge of their interconnectivities. Static knowledge is represented by lists, object-attribute-triplets (OAV), and arrays.

The OAV triplet is extremely useful for the representation of the associated properties of an object. In the OAV triplet representation, objects may be physical or conceptual entities. Attributes are general characteristics or properties associated with objects. Size, shape, and color are typical attributes for physical objects. The final element of the OAV triplet is the value of the attribute. As an example, the properties of a standard steel W-shape such as W36X300 can be represented by OAV representation, as shown in Table 1. The record package of INTERLISP is used to store the properties of 187 W-shapes available in the AISC manual (AISC, 1987). Each record for a W-shape contains an object name (for example, W36X300) and 17 values for 17 attributes, as shown in Table 1.

Dynamic knowledge includes the knowledge of design constraints that have to be satisfied in a given design problem and the heuristics that are used to solve the problem effectively. Dynamic knowledge is represented by production rules and functionals via procedural abstraction. Due to its flexibility, simplicity, and discreteness, the production system provides a powerful way of representing human thought processes (Buchanan and Shortliffe, 1984). Production rules are used to represent design specifications such as the American Institute of Steel Construction (AISC) specifications (AISC, 1980). As an example, Table 2 shows the production system representation of the AISC specifications Section 1.5.1.4.1.

Graphical knowledge represents the structural configuration in graphical images. Graphical knowledge is represented by bitmaps, windows, and menus (Paek and Adeli, 1988).

Structural Design Language

SDL provides a problem solving environment for development of structural design expert systems. SDL has been developed on a Xerox AI

Table 1 Properties of W36x300

Object: W36x300

Attribute	value
Cross sectional area (A)	88.3
Depth (d)	36.74
Web thickness (tw)	0.945
Flange width (bf)	16.655
Flange thickness (tf)	1.68
Distance from outer face of flange to web toe of fillet (k)	2.8125
Radius of gyration of a section comprising the compression flange plus 1/3 of the compression web area, taken about an axis in the plane of the web (rT)	4.39
Ratio of the depth to the flange area (d.Af)	1.31
Moment of inertia about the strong axis (Ix)	20300
Section modulus about the strong axis (Sx)	1030
Radius of gyration about the strong axis (rx)	15.2
Moment of inertia about the weak axis (Iy)	1300
Section modulus about the weak axis (Sy)	156
Radius of gyration about the weak axis (ry)	3.83
Torsional constant (J)	64.2
Plastic modulus about the strong axis (Zx)	1260
Plastic modulus about the weak axis (Zy)	241

machine equipped with Xerox 1108 monitors having a resolution of 1020 by 820. The Xerox AI machine uses 24-bit for a word. It has 1.5 MB of main CPU memory and 42 MB of disk memory. SDL contains six facilities: inference mechanism, explanation facility, debugging facility, matrix manipulation processor, frame design processor, and graphics interface. The inference mechanism in SDL is forward chaining. The debugging facility consists of numerical and geometrical debugging. The geometric debugging function discovers errors in the input data for coordinates of members, in numbering the nodes, and in the formation of the input data (Adeli and Paek, 1988).

SDL provides three types of explanation functions: WHY?, HOW?, and WHAT? The WHY? function provides reasoning about the domain knowledge when the user inquires about the conclusions made by SDL. the HOW? function is used to obtain detailed explanations about rules when the user asks for deeper knowledge about the production rules provided by the WHY? function. In other words, the HOW? function explains the underlying reasoning or theory behind each rule. Finally, the WHAT? function provides explanations about the terminologies and equations used in the production rules. As mentioned earlier, the AISC design specifications (AISC, 1980) have been represented by IF-THEN rules and

Table 2   Production Rules for the AISCS 1.5.1.4.1 Requirements

---

(Rule1 (if ((> Section)
            has flanges continuously connected to the web))
       (then ((< Section)
              satisfies the flanges continuously connected to the web
              requirement)))
(Rule2 (if ((> Section)
            contains unstiffened elements)
           ((< Section)
            has (bf/(2*tf) LT 65/(SQRT Fy))))
       (then ((< Section)
              satisfies flange local buckling requirement)))

. . .

(Rule5 (if ((> Section)
            has (fa/Fy GE .16)))
       (then ((< Section)
              satisfies web local buckling requirement)))

. . .

(Rule7 (if ((> Section)
            is a W-shape)
           ((< Section)
            has (and (Lc LE 76*bf/(SQRT Fy))
                     (Lc LE 20000/(d.Af*Fy)))))
       (then ((< Section)
              satisfies W-shape unsupported length requirement)))

. . .

(Rule12 (if ((> Section)
             satisfies the flanges continuously connected to the web
             requirement)
            ((< Section)
             satisfies flange local buckling requirement)))
            ((< Section)
             satisfies web local buckling requirement)))
            ((< Section)
             satisfies W-shape unsupported length requirement))
        (then ((< Section) is Compact)))

---

functionals. Some of these rules contain long equations. If equations
are used directly in the IF-THEN rules, they become cumbersome and
complicated. Hence, equation numbers are used in the production rules
instead of the equations themselves. However, when the user wishes to
see the equation itself, the WHAT? function displays the equation.

Cooperating Specialists

Brown and Chandrasekaran (1984) present a framework for creation of computer-based expert consultants where knowledge is decomposed into substructures and each substructure is divided into a hierarchy of specialists. They applied this methodology to develop an expert system for mechanical design with design refinement as the central problem solving activity.

In actual design of structures, often a number of senior and junior engineers, technicians, and draftsmen work cooperatively. In order to solve the complex problem of integrated design of structures, a problem solving paradigm has been created based on the concept of cooperating specialists. The complex body of knowledge needed for detailed design of structures is fractionated into smaller and manageable knowledge sources which are then organized into a set of conceptual specialists.

In the case of frame structures, at the topmost level, we have the FrameDesign specialist (Figure 1). This specialist uses the help of four other specialists at the second level of hierarchy. They are the FramePredesign specialist whose task is the preliminary design, the FrameAnalysis specialist whose task is structural analysis, the SectionDesign specialist whose task is the design of individual members (beams and columns), and the ConnectDesign specialist with the task of designing the connections.

The FramePredesign specialist selects preliminary values for the cross-sectional areas and moments of inertia of beams and columns. The hierarchical tree of the FramePredesign specialist is shown in Figure 2. This specialist uses several lower level specialists. For example, the PreSection specialist estimates the beam section modulus and the column cross-sectional area and moment of inertia, using the knowledge of previous designs. This knowledge is extracted from the hundreds of designs given in parts two and three of the AISC manual (AISC, 1980).

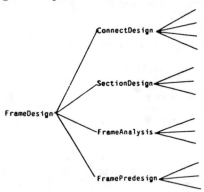

Figure 1 The hierarchical tree of the FrameDesign specialist

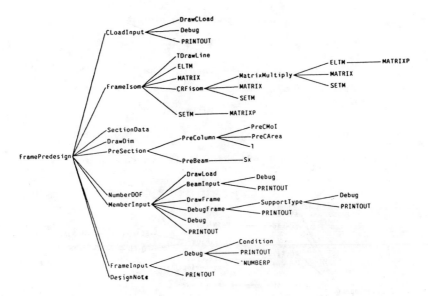

Figure 2   The hierarchical tree of the FramePredesign specialist

The hierarchical tree for the FrameAnalysis specialist is presented in Figure 3. The stiffness method is used for structural analysis. The Cholesky decomposition method is used for the solution of the simultaneous equations.

The hierarchical tree for the SectionDesign specialist is shown in Figure 4. This specialist uses three lower level specialists: SectionInput, CompactSection, and SectionSelect, with hierarchical trees shown in Figures 5 to 7, respectively. The CompactSection specialist checks the compactness of a given section.

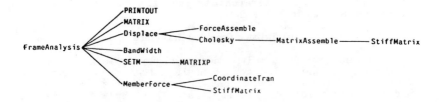

Figure 3   The hierarchical tree of the FrameAnalysis specialist

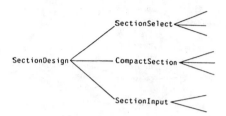

Figure 4   The hierarchical tree of the SectionDesign specialist

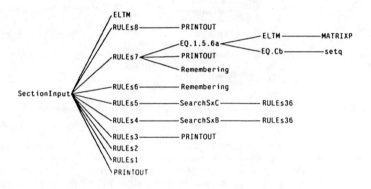

Figure 5   The hierarchical tree of the SectionInput specialist

The ConnectDesign specialist at present designs a shop welded and field bolted moment-resisting connection. This moment connection uses flange plates on both the top and bottom flanges of the beam to transfer the horizontal force caused by the applied moment. The shear force caused by the end reaction is transferred to the column using a single web plate. The hierarchical tree for the ConnectDesign specialist is shown in Figure 8.

Multiwindow Graphics Interface

SDL also provides a multiwindow graphics interface. The graphics interface is capable of displaying both the orthographic and isometric views of the frame structure and beam-column connections (see Figures 9 and 10). An efficient priority list algorithm with back-face elimination has been used for the orthographic views. A five-test algorithm has been employed for efficient drawing of the isometric views. For details of the graphics algorithms, see Adeli and Fiedorek (1986).

INTEGRATED STRUCTURAL DESIGN 47

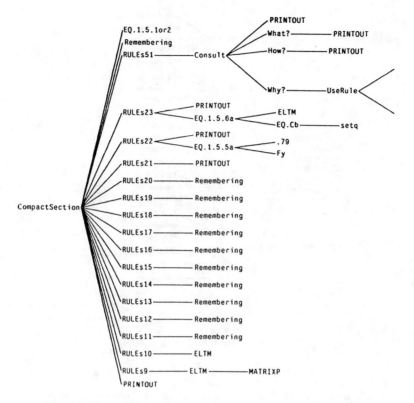

Figure 6 The hierarchical tree of the CompactSection specialist

Application

SDL has been used to develop a prototype expert system for integrated design of steel buildings made of moment-resisting frames, called STEELEX (Adeli and Paek, 1988). STEELEX is a coupled expert system that uses both symbolic and numeric computing techniques. The basis of design is the AISC specification (AISC, 1980). STEELEX can perform
1. preliminary design of members,
2. analysis of the structure,
3. detailed and final design of the structure, including connections, and
4. drawing of the final design.

STEELEX will be presented in detail in a forthcoming article (Adeli and Paek, 1988). Sample multiwindow output is presented in Figures 9 and 10.

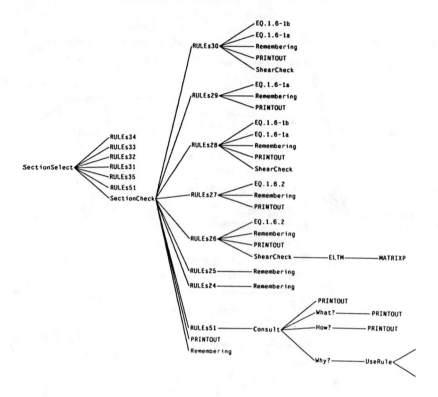

Figure 7 The hierarchical tree of the SectionSelect specialist

It should be pointed that the expert system STEELEX has a hierarchical structure which is fundamentally different from the structure of the conventional CAD programs. The user of the system can request to see the hierarchical tree from any selected node on the screen of the display monitor. The hierarchical tree will be displayed automatically by STEELEX. In fact, the hierarchical tree presented in Figures 1 to 9 have all been generated by STEELEX. Ideally, it is desirable to have a "glass box" program rather than a "black box" program. The major advantage of this hierarchical program structure is its transparency to the user. The user can see inside the program and how it works clearly. The hierarchical program structure also makes the program easier to understand, debug, and maintain.

Another advantage of the new program structure is that different portions of the system (for example, various specialists) can be executed independently. Also, the program can be interrupted by the user at the end of each step and resumed again without any need to start all over again.

INTEGRATED STRUCTURAL DESIGN 49

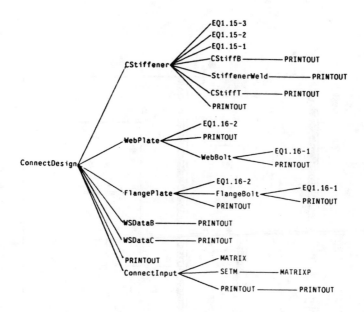

Figure 8. The hierarchical tree of the ConnectDesign specialist

Acknowledgement

The research presented in this paper has been partially supported by the Ohio State University Office of Research and Graduate Studies. This research has been performed at the OSU Laboratory for AI Research.

Appendix – References

Adeli, H., Ed. (1988), Expert Systems in Construction and Structural Engineering, Chapman and Hall, London.

Adeli, H. and Al-Rijleh, M.M. (1987), "A Knowledge-Based Expert System for Design of Roof Trusses", Microcomputers in Civil Engineering, Vol. 2, No. 3, pp. 179-195.

Adeli, H. and Balasubramanyam, K.V. (1988a), "A Knowledge-Based System for Design of Bridge Trusses", Computing in Civil Engineering, ASCE, Vol. 2, No. 1, pp. 1-20.

Adeli, H. and Balasubramanyam, K.V. (1988b), Expert Systems for Structural Design - A New Generation, Prentice-Hall, Englewood Cliffs.

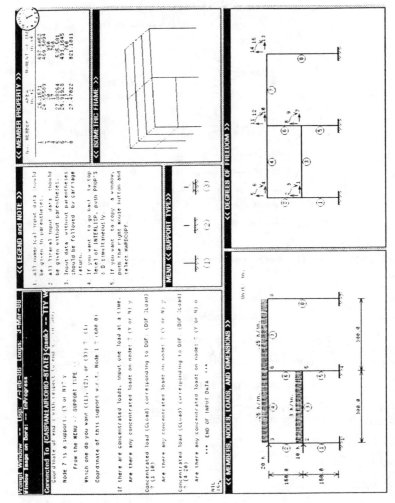

Figure 9 Multi-window graphics for the FramePredesign specialist.

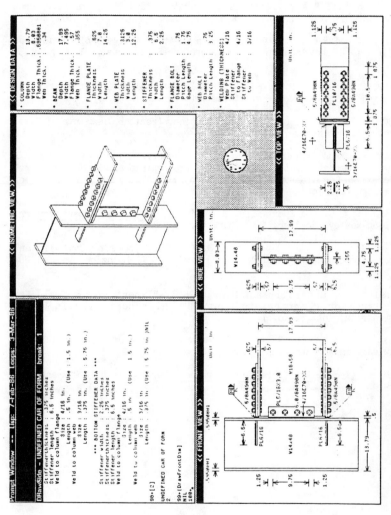

Figure 10 Multi-window graphics for the ConnectDesign specialist

Adeli, H. and Fiedorek, J., "A MICROCAD System for Design of Steel Connections - I - Program Structure and Graphic Algorithms", Computers and Structures, Vol. 24, No. 2, pp. 281-294.

Adeli, H. and Paek, Y. (1986a), "Computer-Aided Design of Structures Using LISP", Computers and Structures, Vol. 22, No. 6, pp. 939-956.

Adeli, H. and Paek, Y. (1986b), "Computer-Aided Analysis of Structures in INTERLISP Environment", Computers and Structures, Vol. 23, No. 3, pp. 393-407.

Adeli, H. and Paek, H. (1988), "STEELEX: A Coupled Expert System for Detailed Design of Steel Structures", to be published.

AISC (1980), Manual of Steel Construction, American Institute of Steel Construction, Chicago, Illinois.

Brown, D.C. and Chandrasekaran, B., (1984), "Expert Systems for a Class of Mechanical Design Activity", Proceedings of the International Federation for Information Processing WG5.2 Working Conference on Knowledge Engineering in Computer-Aided Design, Budapest, Hungary.

Buchanan, B.G. and Shortliffe, E.H. (1984), Rule-Based Expert Systems, Addison-Wesley, Reading, Massachusetts.

Kitzmiller, C.T. and Kowalik, J.S. (1987), "Coupling Symbolic and Numeric Computing in Knowledge-Based Systems", AI Magazine, Vol. 8, No. 2, pp. 85-90.

Moselhi, O. and Nicholas, M.J. (1988), "Expert System Tools for Construction Planning and Control", Microcomputers in Civil Engineering, Vol. 3, No. 1.

Paek, Y. and Adeli, H. (1988), "Representation of Structural Design Knowledge in a Symbolic Language", to be published.

Rich, E. (1983), Artificial Intelligence, McGraw-Hill Book Company, New York.

An Integrated Rule-Based System for Industrial Building Design

B. Kumar[**] and B.H.V. Topping [*]

ABSTRACT

This paper is a brief account of a research project, undertaken in the Department of Civil Engineering at the Edinburgh University, on applying rule-based approaches to the design of industrial buildings. INDEX is a knowledge-based system having a blackboard architecture. The central blackboard is surrounded by ten knowledge modules comprising of heuristics from experts as well as analytical and theoretical knowledge from text-book and other sources. The environment is integrated in that the system commences at the preliminary design stage and goes on to the detailed design of different components of the structure after undertaking a detailed analysis of the chosen structure in the preliminary design stage. The system uses a shell written in the Edinburgh PROLOG called the Edinburgh PROLOG Blackboard Shell. The preliminary analysis routines are written in Edinburgh PROLOG and the detailed structural analysis programs in FORTRAN77. An interface between Edinburgh PROLOG and FORTRAN77 is developed to help maintain the integrated environment.

1. Introduction

The advent of knowledge-based systems in recent years has given an entirely new direction to computer-aided design of structures. There are currently innumerable research projects being actively pursued all over the world. Descriptions of some of the pioneering works can be found in references [1,2,3,4,5]. This paper reports on an ongoing research project concerned with the development of a knowledge-based system for the design of industrial buildings, INDEX. The development of INDEX has been greatly influenced by HI-RISE [6] and DESTINY [7]. These two systems are confined to the domain of residential and commercial buildings.

After a thorough examination of the DESTINY [7] and HI-RISE [6] models, the authors concluded that they were sufficiently general to be used as a sound basis for the development of INDEX. However, due to the difference in the domain, certain modifications had to be made in the domain-specific areas. Although, INDEX is also in the broad domain of building design, there are certain differences in the domain knowledge, e.g., incorporation of cranes is quite common in an industrial building whereas they are non-existent in residential or commercial buildings. In addition, there are also certain differences in the problem-solving approaches adopted by INDEX and HI-RISE as will be discussed in section 3.2.

Due to the comprehensive and general enough models of DESTINY and HI-RISE the main focus of attention in the future has to be in the area of implementation rather than development. This would involve the selection of proper knowledge representation and prolem-solving strategies. Our main efforts also lie in these areas which also include the development of an integrated environment for a complete structural design.

[**] Research student, [*] Lecturer, Department of Civil Engineering, University of Edinburgh, The King's Buildings, Edinburgh EH9 3JL, United Kingdom

2. Some major features and components of INDEX

INDEX has a blackboard architecture. The schematic model of INDEX is given in figure 1. The system is to be utilised after the general layout of the building has been fixed. In other words, the system takes the general layout and other spatial constraints of the building as its input. Since generally the layout of the design is fixed, the domain of the system may be restricted to structural design. In general, the space planning is in the domain of architectural design.

2.1 Blackboard

The blackboard consists of different entries posted on it by the different knowledge modules as the solution of the problem gradually emerges on it. The entries on the blackboard may be seen as a hierarchial decomposition of the industrial building design process.

2.2 Knowledge Base

The levels of knowledge base are given the same names as used by Sriram [7]. However, there are significant differences between the two models. The most important one is the difference in the number of levels of hierarchy in the organisation of the knowledge base. DESTINY's knowledge base is organised into a hierarchy of three levels whereas that of INDEX is in only two. The result is the absence of the Strategy level of DESTINY. The reason for this is that the rules for setting up the Specialist Agenda which the Strategy level module TACON consists are not required in this case. This task is accomplished by giving appropriate 'est' values to the rules as discussed later. The Specialist Agenda for INDEX is :

(ALTSEL->STRANEX->DETEX->OPTEX->EVALUATOR)

This agenda is currently fixed but methods of keeping this agenda flexible according to the situation are being developed. The other difference is the inclusion of the module OPTEX for structural optimisation at the Specialist level and a related module STOPT at the Resource level.

The knowledge base consists of a number of knowledge modules as shown in figure 1. As mentioned earlier, the knowledge modules are organised into a hierarchy of two levels, the specialist level and the resource level. The knowledge modules at the specialist level consist mainly of heuristics and other knowledge that are specialist-dependant. The knowledge modules at the resource level consist mainly of textbook knowledge. All the knowledge modules contain declarative as well as procedural knowledge. A brief description of the knowledge modules at the different levels is given below :

(1) Specialist level : This consists of knowledge modules primarily containing experience-based heuristics. Of course, some textbook knowledge will also be stored at this level.

This level consists of the following knowledge modules :

ALTSEL : This module is responsible for the ALTernative SELection of the feasible structural systems and deciding about different design parametres as the required frame spacing, whether to go for a single or a multi-bay system etc.

STRANEX : This module carries out the modelling and analysis of the chosen structural system by ALTSEL.

DETEX : This carries out the detailed design, i.e. detailed proportioning of the components of the chosen structure.

EVALUATOR : This module evaluates the different alternatives generated by the

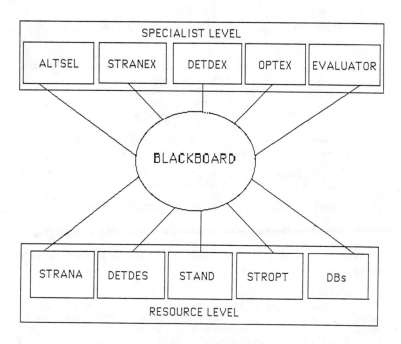

Schematic Model of INDEX

Figure 1

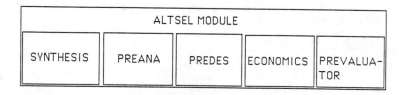

Sub-modules of the ALTSEL module

Figure 2

system.

OPTEX : This module consists of various heuristics and rules to be used for the optimisation of the structures.

(2) Resource level : This level generally consists of algorithmic programs, e.g. structural analysis programs, standard codes, optimisation routines etc.

The knowledge modules at this level consist of the following:

STRANA : This module includes the STRuctural ANAlysis programs.

DETDES : This module is responsible for the DETailed DESign of the structure, i.e. detailed sizing of the components of the structure.

STAND : This module includes the provisions of the applicable STANDards and is responsible for checking these standard constraints.

STOPT : This module consists of STructural OPTimisation routines.

DBs : These DataBases include the different dimensions and sectional properties of various structural sections, e.g. UBs, UCs etc.

Further, these modules consist of a number of sub-modules as for example shown in figure 2 and discussed in section 2.3.

2.3 Control Mechanism

A blackboard shell is being used for the implementation so that the control mechanism is already built into the shell. It consists mainly of an agenda, dynamically built during the consultation process. The agenda sequences the firing of the rules inside a knowledge module. The shell itself is written in the Edinburgh PROLOG [9]. A typical rule in the blackboard shell syntax is of the following form [10,17] :

```
if    Condition
then  Goal
to    Effect
est   Est.
```

The effect of a rule may be one of the following :

add[Index,Fact,Cf], which adds an entry Fact on the blackboard under the index Index with certainty factor Cf,

or amend[Index,Fact,Cf], which amends an entry Fact on the blackboard under the index Index with certainty factor Cf,

or action Action, which takes an action Action,

or delete, which deletes an entry on the blackboard,

where Index, Fact, Cf and Action are PROLOG terms.

The 'est' in the rules indicate the 'usefulness' of each rule and, thus, helps in building up the agenda. So, by giving appropriate 'est' values to the different rules, we may sequence the firing of these rules. The rule with the lowest 'est' value will be fired first. The rule with the next higher conflict-resolution provided by the shell by giving numerical values to this approach and we have defined a strategy that suits the requirements of INDEX. As mentioned earlier, the sequence of execution of the different knowledge-modules is as follows :

(ALTSEL -> STRANEX -> DETDEX -> OPTEX -> EVALUATOR)

Each of these modules consists of different sub-modules as shown in figure 2. The ALTSEL module will be used to explain our conflict-resolution strategy. The sequence of execution of the sub-modules of ALTSEL is as follows :

(SYNTHESIS -> PREANA -> PREDES -> ECONOMICS -> PREVALUATOR)

One approach to do it is to simply give numerical values to 'est' starting from the first rule of SYNTHESIS and increment them upto the last rule of PREVALUATOR. This, obviously, is not an elegant approach to the problem since if one adds a rule to any of the modules at a later stage, all the 'est' values have to be changed for all the rules following it. Another reason is that the whole idea of modularity gets lost by this approach and the set of rules, in effect, becomes one module instead of being broken down into sub-modules.

The second approach adopted is to give symbolic 'est' values to the rules of the different sub-modules specific to the rules of that sub-module only and define a different conflict-resolution strategy altogether. In this approach, the sequence of execution of the sub-modules is first defined and then the sequence of firing the rules inside each sub-module. The following example is a rule from the SYNTHESIS sub-module. This approach avoids the problems of the default approach described above.

```
if    [problem,span(X),true]
and   holds(X =< 60)
then  output_message('Single span portal frame is a feasible alternative')
to    add[synthesis,lateral_load_sys(single_span_portal),true]
est   synthesis(5).
```

The 'est' value of this rule indicates that this rule is the fifth rule inside the SYNTHESIS sub-module. The numerical value in this 'est' decides the firing of the rule inside that sub-module and the invoking of the sub-modules is decided by the top level conflict-resolution.

One important drawback of using a shell of this type is that developing a multi-formalism system becomes practically infeasible. For example, the shell used in INDEX is a forward-chaining production rule system and any need to use a different formalism, like frames for example, would present a formidable task. One notable feature of this shell, however, is that one may easily switch to a backward-chaining formalism by writing the relevant portions in PROLOG. It is worthwhile emphasising here that using this shell has not proved to be a handicap at all so far and incorporating some additional important features into the system has proved to be quite simple. For example, setting up the Specialist agenda or incorporating the ability to change this agenda at the user's discretion have already been found to be quite straightforward to implement without requiring a separate set of production rules as in DESTINY.

2.4 Explanation facilities

The system also possesses some explanation facilities. Work is still being carried out to improve the existing explanation facilities. Basically, two approaches are being investigated. One is to have an associated explanation with every conclusion the system may arrive at. The other approach is to use the front-end facilities of the shell and generate expalnations using them. Some of the front-end commands of the shell are :

show rules - this prints out all the rules assimilated in the system.

show next - this gives an account of the details of the next step to be taken.

show this - this gives the details of the current step.

show agenda - this gives the agenda under consideration in the present cycle.

show supports_of(X) - this gives the supports of the entry X.

step - this gives a summary of the last cycle.

show rule(N) - this gives the rule number N.

These are only a few of the front-end facilities provided by the shell. At the moment, the explanations that may be obtained from the system:

1. the rule or set of rules that forced a particular conclusion;

2. the current entries on the blackboard;

3. the reasons for concluding something;

4. the set of rules that were successful at the end of a session; and

5. the details of any of the alternative feasible solutions generated by the system.

The examples given in Appendix I will illustrate these explanation facilities of the system.

3. Implementation

An initial implementation of the ALTSEL module has been undertaken incorporating approximately one hundred rules. It is being implemented on a Sun 3/50 workstation. The system has knowledge about the following types of steel frames :

1. portal frames;

2. roof trusses and columns; and

3. beams and columns.

Apart from these, it also has rules for incorporating gantries for the design of gantry cranes if required. The solution search space for the feasible lateral load resisting systems is shown in figure 3. The search strategy adopted is the breadth-first search. The system generates all the solutions at one level before going on to the next level. In the current approach, the system is solitary and not interactive. It takes some input and works through the knowledge base printing out the different solutions at every stage. The sequence in which it searches its knowledge base and prints out the solutions is as follows :

1. all the feasible lateral load resisting frames,

2. economic frame spacing and type of purlin,

3. alternatives for sides,

4. alternatives for side-cladding,

INDUSTRIAL BUILDING DESIGN

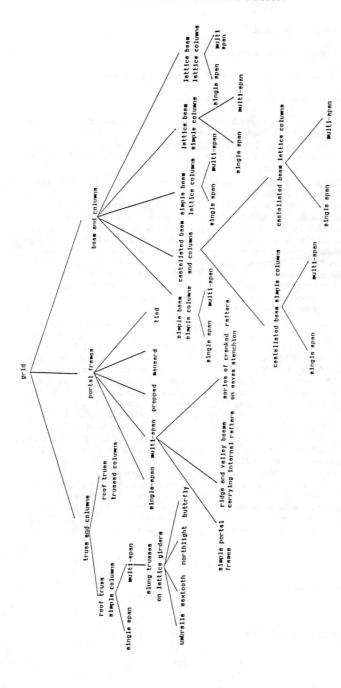

Figure 3. Solution search space for the feasible lateral load resisting systems

5. alternatives for roof system

6. approximate analysis of the lateral load resisting frames,

7. approximate sizing of the feasible lateral load resisting frames and

8. relative evaluation of the feasible lateral load resisting frames.

One important drawback of this approach has been the lack of transparency of the system. The user does not get the complete details of a particular alternative at a glance. To overcome this drawback, the user is given the facility of obtaining the complete details of any alternative solution generated by ALTSEL at the end of the session using the 'show_details_of' command (see appendix I).

The ALTSEL module has five sub-modules as shown in figure 2, viz., SYNTHESIS, PREANA, PREDES, ECONOMICS AND PREVALUATOR. Based on the rules in these sub-modules, the system is able to select the feasible alternatives for the lateral load resisting main frames for the industrial building in question. An approximate analysis is undertaken by the PREANA sub-module for each of the alternatives generated by the SYNTHESIS sub-module. The plastic method of analysis is used for this approximate analysis except for the lattice beam and truss roof altrernatives. An approximate proportioning and sizing of all the alternatives is then undertaken by the PREDES sub-module. These sizes are picked up from a database containing Universal Beams, Universal Columns and Square Hollow sections. The relative economics of the alternatives generated are then considered by the ECONOMICS sub-module. The rules in this sub-module are based on a study by Morris and Horridge [11] on the comparative costs of single-storey steel framed structures.

Considerable effort was made to calibrate rules using experts and design case studies provided by them. Although some of the rules are based on discussions with working design engineers [12], most of them are taken from published literature from various steel section and frame manufacturing and fabricating firms and organisations such as the Steel Construction Institute (formerly known as the Constructional Steel Research and Development Organisation) [13]. However, the system has proved to be quite effective on some real-life problems which have been used to test INDEX. A sample run of the system is given in Appendix II. The problem was to design a factory building of 55 metres span to withstand a vertical load of 7.5 tons/m. Other inputs were to find out any relevant detailed design constraints. No internal stanchions were allowed inside the building. By following the list of entries on the blackboard and a run, one may understand the sequence of the emergence of the solution on the blackboard. One difference between DESTINY and INDEX here is that whereas DESTINY only posts the current best design (CBD) on the blackboard, INDEX posts all of them. This is done so as to provide greater flexibility in terms of being able to review the whole context tree at a glance. An entry on the blackboard is identified by the index attached to it. In the example given below, the index of each entry is 'synthesis', their status is 'in' which means that they have been added into the list of entries on the blackboard, their certainty is 'true' and the 'facts' added are a list of lateral load system alternatives, viz., the lattice beam, the castellated beam and the tied portal. The numbers at the begining of these entries are not relevant in the present context.

```
114 in true
synthesis
lateral_load_sys(single_span_portal)

325 in true
synthesis
lateral_load_sys(beam_and_columns)
```

142 in true
synthesis
lateral_load_sys(tied_portal)

3.1 Representative production rules

Much work has already been done on the ALTSEL module of the specialist level. This module is responsible for the preliminary design of the building, e.g. selecting the different alternative structural systems for the building, selecting the appropriate frame spacings, selecting the most economical system, etc. ALTSEL takes the column grid layout of the proposed building as its input. This module consists mainly of heuristics obtained from different design literature and some practicing engineers. This is a simple representative rule from this module:

    if   [synthesis,span(X),true]
    and   [problem,int_stanch(yes),true]
    and   holds(X > 20)
    then output_message('Propped portal is an alternative.')
    to   add[synthesis,lateral_load_sys(propped_portal),true]
    est   synthesis(3).

which means that, if the span is known and is less than or equal to sixty metres, then a single span portal is a feasible alternative; 'output_message' is a simple PROLOG procedure.

3.2 Problem-solving strategy

All the different problem-solving strategies in Artificial Intelligence adopt one of the following two approaches [18] :

1. Formation approach and

2. Derivation approach.

The formation approach involves the formation of the most appropriate solution by putting together the different components of a complete solution stored in the knowledge-base at different levels. In contrast, the derivation approach involves picking up the most appropriate solution from a set of pre-defined solutions already stored in the knowledge-base.

It is quite evident that the formation approach is probably more general and intelligent way of solving a problem. However, for the domain we are working in, we found that the derivation approach provided an easier way of solving the problems. This was decided after experimenting with the formation approach. Hence, INDEX utilises the derivation approach to solving the problem. This is in contast to HI-RISE which utilises a formation approach [18]. Figure 4 is an inference network for the selection of feasible lateral load systems. The knowledge-bases of INDEX consist of different feasible solutions for different situations. In doing so, it proceeds ahead by handling different constraints [14], which consists of the following :

1. constraint formulation,

2. constraint satisfaction and

3. constraint posting.

This concept of constraint handling is accomplished in the system by first *satisfying the constraint* for a particular alternative, *looking for any constraint* associated with the alternative which will be used by other modules, i.e., constraint formulation, and *posting* it to the

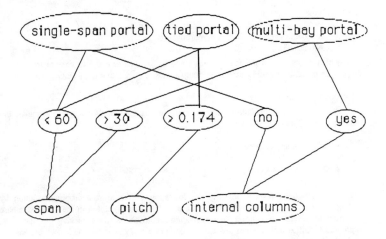

Figure 4. Inference network for lateral load resisting frames

appropriate module which is supposed to use it later on, i.e, constraint posting. This may be illustrated by the following rule.

    if   [synthesis,lateral_load_sys(multi_portal),true]
    then output_message('The following constraints should be taken into account in the detailed design stage of multi-bay portals with cranked rafters :-
         a)spread under load could be critical and should be examined carefully.')
    to   add[design,design_cons(multi_portal),true]
    est   des(6).

In this example, the constraint has been posted on the blackboard with the index 'design'. It may be accessed by any other module by this index. For example, the following rule from the standard provision checking module states that, if there is an entry on the blackboard that says that there are some constraints to be satisfied in the detailed design of multi-span portal frame then invoke the PROLOG procedure that tests whether those constraints are satisfied.

    if   [design,design_cons(multi_portal),true]
    then check_multi_portal_cons
    to   add[design_check,multi_portal_cons(satisfied),true]
    est   des(12).

where 'check_multi_portal_cons' is a PROLOG procedure that tests whether the constraints are satisfied or not.

3.3 Types of constraints

The different types of constraints considered by different knowledge modules are different and depend upon the task being performed. The constraints considered by the sub-modules of the ALTSEL module are mostly external. For a comprehensive description of different types of constraints in structural design, readers are directed to [8]. External constraints are the constraints that are not in the hands of the designer. In other words, these constraints are *external* to the designer. These constraints are mostly governed by the requirements of the client. The constraints considered by the SYNTHESIS sub-module in deciding about the feasible lateral load resisting systems are shown in figure 5. Figure 6 shows the considerations used by the PREVALUATOR sub-module in evaluating the different feasible systems.

4. The Interface

In order to maintain the integrated environment of INDEX an interface between FORTRAN and Edinburgh PROLOG was developed. Two interfacing approaches were explored as discussed below.

(1) The file approach - this is a simple way of developing an interface. In this approach, the knowledge-based component and FORTRAN communicate via files. Some implementations of PROLOG (e.g., Quintus-PROLOG, Edinburgh PROLOG and C-PROLOG) allow for calling the operating system and executing any system command from PROLOG. This facility is the key to this method. The knowledge-based component stores all the input data for the FORTRAN program in a file. The FORTRAN program is invoked by a system call from PROLOG. Similarly, the FORTRAN program stores its output on a file which is read from PROLOG. This approach is quite straightforward. However, it does not provide a fully integrated environment.

(2) The C interface approach - with this approach, the communication between PROLOG and FORTRAN is through C functions as the intermediary. Some implementations of PROLOG

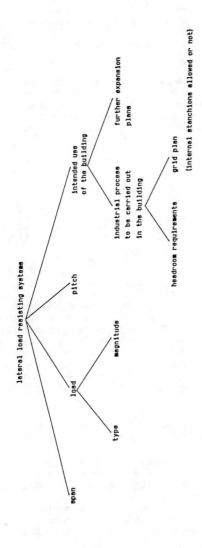

Figure 5. Types of constraints considered by the SYNTHESIS sub-module for the selection of feasible lateral load resisting systems

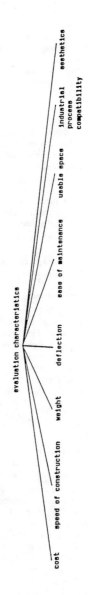

Figure 6. Evaluation characteristics considered by the PREVALUATOR sub-module

INDUSTRIAL BUILDING DESIGN

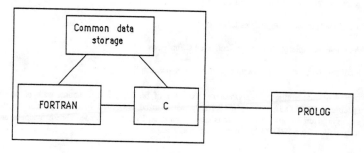

The Interface

Figure 7

$$[A] = \begin{bmatrix} A11 & A12 & A13 \\ A21 & A22 & A23 \\ A31 & A32 & A33 \end{bmatrix} = [[A11,A12,A13],[A21,A22,A23],[A31,A32,A33]]$$

Representation of Arrays in PROLOG

Figure 8

(e.g., Quintus-PROLOG, Edinburgh PROLOG) allow for loading of pre-compiled object codes. On the other hand, the Berkley Unix implementation of FORTRAN provides for calling pre-compiled FORTRAN function from C and vice-versa. Thus, the strategy adopted for INDEX was to call C from PROLOG and then FORTRAN from C . To accomplish this, both the C and FORTRAN functions are compiled and linked together in one file. This linked file is, then, loaded in Edinburgh PROLOG. Since the compiled FORTRAN and C functions are linked together, both share common data storage (see figure 7). Thus, FORTRAN functions have direct access to any data passed from PROLOG to C. Similarly, any output data from FORTRAN is directly accessed and sent to PROLOG by C.

Pre-compiled codes may be loaded in Edinburgh PROLOG [16] by the load/3 predicate, where 3 stands for the number of its arguments, as given below :

load(ListofPredSpec,ObjectFiles,Libraries).

A simple example for loading pre-compiled C functions is :

load([pred/1=cfunc1, pred/2=cfunc2],'tmp.o','-lm -termcap').

The first argument *ListofPredSpec* is a list of predicate specifications, each of which specifies a PROLOG predicate and its arity that is to be associated with a C function. In the example given above two C functions *cfunc1* and *cfunc2* are to be loaded. These functions are to be called from PROLOG by pred/1 and pred/2 respectively.

4.1 Accessing arrays

The communication of simple variables does not prove any problem but accessing arrays requires further consideration. The arrays are represented as lists in PROLOG. One-dimensional arrays are represented as single lists whereas multi-dimensional ones are represented as lists containing sub-lists as shown in figure 8. Each sub-list in figure 8 represents a row of the equivalent C two-dimensional array. The array that is passed back from C is also a list consisting of sub-lists. The methods of accessing these arrays in the two methods of interfacing are described below.

(1) The file approach - The elements of the lists representing the arrays in PROLOG are written to a file which is then read by the FORTRAN program. The FORTRAN program writes its output on a file. The output arrays are then read and formed into lists by PROLOG. Each element of the arrays is passed back to PROLOG one by one and the corresponding list keeps getting constructed in PROLOG using a simple PROLOG procedure .

(2) The C interface approach - The elements of the lists representing the arrays in PROLOG are passed one by one recursively to FORTRAN via C. Similarly, the elements of the output arrays from FORTRAN are passed back one by one and the corresponding list gets constructed recursively in PROLOG .

Considering the mapping of arrays between C and FORTRAN, the most important requirement is that the dimensions of the arrays have to be altered. This is because FORTRAN 77 arrays are stored in column-major order whereas C arrays are stored in row-major order. Thus, the column dimension of a FORTRAN 77 array will become the row dimension of the equivalent C array and the row dimension of FORTRAN 77 array will become the column dimension of the equivalent C array. For example, an array A(3,2) in FORTRAN will have to be represented as A[2][3] in C and will be the transpose of the actual array required in FORTRAN. Another important point is that the first element of a C array always has a subscript zero whereas FORTRAN 77 array elements always begin with a subscript of one.

Details of tests undertaken to assess the efficiency of the methods for passing and returning arrays may be found in reference [15]. It is clear that the C interface approach is almost three to four times faster than the file approach, on an average. It was also concluded that the choice between the two approaches will become crucial only when the number and sizes of the arrays to be used by the FORTRAN program are quite large. For example, if a program uses 100 arrays of 1000 elements each, the difference in time in just passing these arrays would be almost 7 minutes. Although, this difference would not be significant for a program that runs for hours, it is quite considerable. On the other hand, for graphics and other user interface facilities, time is of utmost importance and the obvious choice in that case would be the faster approach even if the time difference is minimal.

5. Conclusions

The following conclusions were drawn when undertaking the research reported here :

1. Artificial Intelligence tools and techniques, particularly expert system shells, provide an easy way of incorporating rules of thumb, heuristics etc. in computer programs with minimum effort.

2. In the preliminary design stage, a lot of decisions are taken using rules-of-thumb. Thus, a system like INDEX might perform satisfactorily in that domain.

3. In the domain of structural design, the shell or language used in the development of the system should have the capability to interface with procedural structural analysis programs.

4. The main problem in developing an integrated knowledge-based system for the design of structures is the inability of the AI languages like LISP and PROLOG or expert system shells to interface with procedural languages like FORTRAN (although some shells like OPS83 are exceptions).

5. The main obstacle in developing an interface between FORTRAN and PROLOG was the communication of arrays. A methodology was suggested to overcome this problem.

6. Acknowledgements

The authors acknowledge the collaboration with the Engineers of Babtie Shaw and Morton, Glasgow and Ove Arup and Partners, Edinburgh in the implementation of INDEX. The research described in this paper is part of a larger collaborative project being undertaken by the Department of Civil Engineering and the Artificial Intelligence Applications Institute at the University of Edinburgh. Bimal Kumar wishes to acknowledge the University of Edinburgh for his postgraduate research scholarship. The authors also acknowledge with thanks the help and suport extended by Robert Rae and Paul Chung of the Artificial Intelligence Applications Institute, University of Edinburgh.

References

1. Sriram, D., Maher, M.L., Bielak, J. and Fenves, S.J., *Expert Systems for Civil Engineering - A Survey*. Report No. R-82-137, Department of Civil Engineering, Carnegie-Mellon University, U.S.A., July '82.

2. Sriram, D., Maher, M.L. and Fenves, S.J., *Knowledge-Based Expert Systems in Structural Design*. Computers and Structures, Vol. 20, No.1-3, pp. 1-9, 1985.

3. Adeli, H., *Knowledge-Based Systems in Structural Engineering* in The Application of Artificial Intelligence Techniques to Civil and Structural Engineering, Ed., Topping, B.H.V., pp.71-78, Civil-Comp Press, Edinburgh, 1987.

4. Kumar, B., Chung, P.W.H., Rae, R.H. and Topping, B.H.V., *A Knowledge-Based Approach to Structural Design* in The Application of Artificial Intelligence Techniques to Civil and Structural Engineering, pp.79-92, Civil-Comp Press, Edinburgh, 1987.
5. Topping, B.H.V. (editor), *The Application of Artificial Intelligence Techniques to Civil and Structural Engineering*, Civil-Comp Press, Edinburgh, 1987.
6. Maher, M.L. and Fenves, S.J., *HI-RISE: A Knowledge-Based Expert System for the Preliminary Design of High Rise Buildings*. Report No. R-85-146, Department of Civil Engineering, Carnegie-Mellon University, U.S.A., 1985.
7. Sriram, D., *DESTINY: A Model for Integrated Structural Design*. Artificial Intelligence in Engineering, Vol.1, No. 2,pp. 109-116, 1986.
8. Sriram, D., *Knowledge-Based Approaches to Structural Design*. Ph.D. Dissertation, Department of Civil Engineering, Carnegie-Mellon University, U.S.A., 1986.
9. Artificial Intelligence Applications Institute, University of Edinburgh, *Edinburgh PROLOG (The New Implementation) User's Manual, Version 1.5*. 1987.
10. Chan, N. and Johnson, K., *Edinburgh Blackboard Shell User's Manual*. Artificial Intelligence Applications Institute, University of Edinburgh, 1987.
11. Horridge, J.F. and Morris, L.J., *Comparative Costs of Single storey Steel Framed Structures*. The Structural Engineer, Vol. 64A, No. 7, pp. 177-181, 1986.
12. Chung, P.W.H. and Kumar, B., *Knowledge Elicitation Methods: A Case Study in Structural Design* in The Application of Artificial Intelligence Techniques to Civil and Structural Engineering, Ed., Topping,B.H.V., Civil-Comp Press, Edinburgh, pp. 21-26, 1987.
13. Bates, W., *Introduction to the Design of Industrial Buildings*. Constrado, Croydon, 1978.
14. Stefik, M., *Planning with Constraints - MOLGEN Part 1*. Artificial Intelligence, No. 16, pp. 111-140, 1981.
15. Kumar, B., Chung, P.W.H. and Topping, B.H.V., *Approaches to FORTRAN-PROLOG interfacing for an Expert System Environment* in The Application of Artificial Intelligence Techniques to Civil and Structural Engineering, Ed., Topping, B.H.V., Civil-Comp Press, Edinburgh, pp. 15-20, 1987.
16. Artificial Intelligence Applications Institute, University of Edinburgh, *Edinburgh PROLOG (The New Implementation) User's Manual, Version 1.4*, 1986.
17. Jones, J. and Millington, M., *An Edinburgh PROLOG Blackboard Shell*, Department of Artificial Intelligence, University of Edinburgh, 1986.
18. Maher, M. L., *Problem-Solving Using Expert System Techniques*, Expert Systems in Civil Engineering, Eds., Kostem, C.L. and Maher, M.L., ASCE, New York, pp. 7-17, 1986.

INDUSTRIAL BUILDING DESIGN

Appendix I

Explanation facilities

```
| ?- start,run.
```

% the user querries are preceded by double asterisk (**) and
% the system output by a single asterisk (*)

* single-span portal lateral load system is possible

    ** why.

* [[205,problem,span(45),true],[210,problem,int_stanch(no),true]]

* tied portal lateral load system is possible

    ** rule.

```
*     if    [problem,pitch(_128807),true]
            and holds _128807>0.174
      then  output_message(tied portal lateral load system is possible)
      to    add [synthesis,lateral_load_sys(tied_portal),true]
      est   synthesis(4)
```

* The following alternatives of multi-bay portal are also feasible :-
        1. Heavy portals at convenient spacings with ridge and valley beams carrying intermediate rafters,
        2. Series of cranked rafters carried on eaves stanchion and valley beams.

    ** entries.

* Blackboard entries as
        tag, status, certainty
        index
        fact :

224 in true
synthesis
lateral_load_sys(single_span_portal)

222 in true
problem
find_lateral_load_sys

216 in true
problem
cr_surge_mom(57.5)

214 in true
problem
crane_base_dist(5.2)

213 in true
problem
bases(fixed)

212 in true
problem
load(5.43)

211 in true
problem
find_design_cons(yes)

```
210  in   true
problem
int_stanch(no)

209  in   true
problem
apex_ht(3.75)

208  in   true
problem
eaves_ht(7.6)

207  in   true
problem
pitch(0.33)

205  in   true
problem
span(45)

yes
| ?- show_details_of(single_span_portal).

Following are thw details of the single span portal alternative :-

ssp_stanchion_sec   %   stanchion section type

ub

ssp_rafters_sec     %   rafters section type

ub

ssp_zp_provided     %   plastic modulus provided

2362

ssp_feas_sec_raft_stanch   % feasible rafters and stanchion
                           % sections

533x210UB@92


yes
| ?- ^D
Prolog terminated
```

INDUSTRIAL BUILDING DESIGN

Appendix II

A sample run of the system

| ?- start,run.

% search for the feasible lateral load frames
single-span portal lateral load system is possible

tied portal lateral load system is possible

Multi bay latticed girder lateral load system is possible

The following alternatives of multi-bay portal are also feasible :-
    1. Heavy portals at convenient spacings with ridge and valley beams carrying intermediate rafters,
    2. Series of cranked rafters carried on eaves stanchion and valley beams.

% advice on the frame spacing and the type of purlins

The frame spacing for the span under consideration should be in the range of 10 to 12 metres and lattice purlins would prove economical. However, in this case, an intermediate frame becomes a necessity between two main frames. Type in a desired value followed by a full-stop.   10.

```
*************************************************************
The following alternatives can be considered for
the sides :-
*************************************************************
```

One alternative is to just have side rails attached to the side stanchions.

Another alternative could be have side bracings between the stanchions.

```
*************************************************************
The following alternatives can be considered for the
side cladding :-
*************************************************************
```

One alternative for side cladding is to have plastic coated sheeting all over.

brickwork in one of the following ways :-
    1. supported from the structure both vertically and horizontally,
    2. supported only horizontally,
    3. self_supporting both vertically and horizontally,
    4. self_supporting and also supporting some elements such as the ends of purlins at the gables.

Another alternative for the side cladding is to have precast or cast-in-situ concrete wall all over.

```
*************************************************************
The following alternatives can be considered for
the roof system :-
*************************************************************

One alternative for the roofing system is to have
cladding simply over purlins.


Another alternative for the roofing system is to have
bracings between the rafters of the supporting frame.


*************************************************************
Following are the approximate section sizes for the
different alternatives of feasible lateral load systems :-
*************************************************************

Following are the feasible sections
for the single span portal alternative :-

686x254UB@140
Zp provided = 4552



The following are the feasible sections for the tied
portal rafters and columns based on aproximate analysis :-

610x178UB@91
Zp provided = 2484

Following is the dimensions for the tie based
on approximate analysis :-

60x60x10 angle or a rod of 36mm. dia.



*************************************************************
The following design constraints should be considered
in the detailed design stage :-
*************************************************************
The following things should be considered in the
detailed design stage of single span portal alternative :-.
                        1.pitch should be kept low because greater slope
will give rise to greater spread at knees which can cause problems with
cladding,
                        2.horizontal thrusts should be carefully examined
and the foundation designed accordingly,
                        3.haunch should be provided at
the eaves and the ridge should be deepened because the maximium bending
moment will occur at the knees.


The following constraints should be taken into account
in the detailed design stage of multi-bay portals with cranked rafters :-
                        a)spread under load could be critical and should be
examined carefully.


yes
| ?- ^D
Prolog terminated
```

A Rule-Based System for Estimating Snow Loads on Roofs

P. Fazio[*], M.ASCE., C. Bedard[**] and K. Gowri[***]

ABSTRACT

The intensity of snow loads on roofs according to the National Building Code of Canada is obtained by multiplying the specified ground snow load with a series of adjustment factors. The intensity of ground snow load for about 480 locations across Canada is provided by the Supplement to the Code. These values are based on observations and computations for a thirty-year return period. The adjustment factors are dependent on the roof type, roof slope, and accumulation characteristics. The Code specifies the conditions, formulas, etc. to be used in calculating each of the adjustment factors. Rule-based systems provide a knowledge representation methodology which is suitable for automating Code specifications. This paper presents the implementation details of a rule-based system which is interfaced to a data base of ground snow loads. The Code clauses required for evaluating the adjustment factors and in estimating the snow loads on roofs are encoded in the rule base. The ground snow load for 120 locations are stored in a data base. The rule base and data base are independent and can be incrementally developed, updated and maintained. During the consultation process, the rule base accesses the data base and obtains the required ground snow load for any given location. This approach is suitable for automating Code specifications which require interface to a large data base.

INTRODUCTION

The potential of knowledge-based expert system (KBES) techniques for solving engineering design problems has been well recognized, and attempts are being made to develop design assistants in architectural, structural and mechanical engineering (1, 2 and 3). The concept of knowledge-based expert system, its components and technical details have been reported by Fenves, Gero and others (1,4). Earlier attempts to develop applications of KBES were faced with the non availability of suitable software tools other than AI programming languages LISP and PROLOG. Over the past few years, a number of KBES development tools have been developed and marketed, many of them to operate on personal computers. The characteristics of these tools vary depending on the knowledge representation methodology, inferencing technique, user friendliness of the development-user interface and the development of end-user interface. Current work at the Centre for Building Studies has concentrated on the evaluation of available development tools and their suitability to solve building design problems. This paper presents the implementation methodology for a snow load estimation system using GURU, a rule-based system development tool.

---

[*]   Director and Professor, Centre for Building Studies, Concordia University, Montreal, Canada H3G 1M8
[**]   Assistant Professor, Centre for Building Studies, Concordia University, Montreal, Canada H3G 1M8
[***] Project Engineer, SIRICON INC., Centre for Building Studies, Concordia University, Montreal, Canada H3G 1M8

Production rules are the most popular form of knowledge representation methodology used in the development of knowledge-based systems. Production rules generally consist of a set of conditions and actions. The advantage of using production rules over the conventional programming methodologies can be realized when a program control flow is very complex or when a program is expected to be modified over a long period of time.

An important aspect of building design in which knowledge-based expert systems can be successfully applied is Code compliance checking, by automating the Code specifications as part of the design assistance offered to engineers and architects. Code specifications are a collection of facts and experiential knowledge which provide mandatory and suggested requirements in designing the various components of a building. One of the major problems of Code automation is to foresee all the possible cross referencing. This can be solved with production systems, where the rules will have all the necessary conditions for cross referencing and the inference mechanism will control the execution process. In the present investigation a system for estimating the snow loads on roofs according to the National Building Code of Canada (5) is developed. A data base of ground snow loads for various locations as specified by the Supplement to the Code (6), is interfaced to a rule base of Code clauses for evaluating the adjustment factors in order to estimate the roof snow loads. The details of the snow load estimation procedure and the implementation details of the system are described in the following sections.

SNOW LOADS ON ROOFS

Part 4 of the National Building Code of Canada (NBCC) deals with the structural design requirements for buildings and members including formwork and falsework. This part of the Code also specifies the procedures to determine the various types of structural loads acting on a building during its life span. Subsection 4.1.7 consists of the details for estimating the intensity of load due to snow, ice and rain. This subsection consists of three articles having a total of ten sentences, eleven clauses and three subclauses. Fig.1. shows the first sentence of the article 4.1.7.1 from the Code.

4.1.7.1.(1) The specified loading, S, due to snow accumulation on a roof or any other *building* surface subject to snow accumulation shall be calculated from the formula

$$S = S_o \cdot C_b \cdot C_w \cdot C_s \cdot C_a$$

where
S_o is the ground snow load in kPa, determined in accordance with Subsection 2.2.1.,
C_b is the basic roof snow load factor of 0.8,
C_w is the wind exposure factor in Sentence (2),
C_s is the slope factor in Sentence (4), and
C_a is the accumulation factor in Sentence (5).

Fig.1. Sentence [1] of article 4.1.7.1 from NBCC(5)

The minimum design snow load on a roof area is obtained by multiplying the ground snow load by the adjustment factors which depend on the wind exposure, slope of roof, roof type and its accumulation characteristics. The Supplement to the National Building Code of Canada (6) provides the climatic information on the intensity of ground snow loads (S_0) for about 480 locations across Canada. These ground snow load values have a thirty year return period and are based on observations and data collected at various stations in the country.

(4) The slope factor, C_s, shall be
  (a) 1.0 when the roof slope, α, is equal to or less than 30°,
  (b) $1.0 - \left(\dfrac{\alpha - 30°}{40°}\right)$ when α is greater than 30°,
      but not greater than 70°,
  (c) 0 when α exceeds 70°, and
  (d) 1.0 when used in conjunction with accumulation factors for increased snow load as given in Clauses (5)(b)(ii) and (5)(b)(v). (See Appendix A.)

Fig.2. Sentence [4] of the article 4.1.7.1 of NBCC(5)

The slope factor (C_s) is dependent on the angle of inclination of the roof to the horizontal. Fig.2. shows sentence [4] of article 4.1.7.1 used to find the slope factor. Eventhough the estimation of slope factor seems easier, it requires a careful interpretation as the reference to subclauses (5)(b)(ii) and (5)(b)(v) is dependent on the roof type. The details of valley type roofs and roofs which may by affected by sliding of snow from an adjacent roof, which are affected by clause 4(d), are given in the Supplement. The accumulation factor (C_a) is dependent on the roof shapes. The common roof shapes such as shed, gable, arch and valley roofs are dealt with in the Commentary H of the supplement to the Code. All roof snow loads estimated must be applied as uniformly distributed loads over the entire area. For flat, shed, low sloped gable and shallow arch roofs, two loading cases have to be considered. One load case with full distributed loading and the other load case with half of the roof with half the intensity of the calculated loading. More load cases and different intensities of loading must be calculated for the arch roofs, valley roofs and multi-level roofs. All the various load cases are necessary to consider the load distribution that may result from drifting, scouring and sliding of the snow. Detailed discussions of snow load estimation for various types of roofs can be found in ref.7 and 8.

A simple rule form can be used to represent the knowledge required in calculating the adjustment factors. By classifying the roof types and the accumulation characteristics, the required number of load cases and corresponding details can be found. The ground snow loads can be obtained from the climatic data knowing the location of the building. It is desirable to maintain a separate data base of the ground snow loads, so that the updating and maintenance can be independent of the rules.

FEATURES OF THE DEVELOPMENT TOOL

In the initial stages of this study, M.1, one of the first knowledge-based system development tool on personal computer was considered for implementation. The Code clauses of article 4.1.7.1 corresponding to the estimation of snow loads on flat, shed and gable type roofs were represented as M.1 rules. It was found that a great number of facts were required to specify the ground snow loads for the various locations. This was undesirable and the expected system would be too large and inefficient. Even though M.1 provides a good tracing facility for debugging the rule-base, it was found to be not suitable for the snow load estimation problem.

The development tool used in the implementation reported here is Guru (10), which is a knowledge-based expert system development tool for the business application software market. Guru provides a development environment which integrates the capabilities of a rule-based system, data base management system, spreadsheet analysis, business graphics, text processing and procedural language programming. The advantage of this development system is the flexibility due to integration and the versatility of its components.

For example, the premise of a rule may consist of reference to data base fields, spread sheet cells, program variables, etc. Knowledge-based system development in Guru proceeds as follows: definition of a goal, initialization phase, definition of environment and program variables, rule base and completion phase. A menu interface is provided for the development-user to use templates for creating the rule base, data base, variable definitions, etc. Alternatively, a normal text editor can be used in part of the development. Once the development of the rule base and the necessary components are complete, a compilation must be done before doing any consultation with the knowledge base. Such a strategy enables the protection of knowledge base in an encrypted form and can be commercially distributed with a run-time version of the Guru system.

IMPLEMENTATION DETAILS

In calculating the roof snow load, the basic data needed is the ground snow load for a given location. The data mangement scheme in Guru is used to create a data table of city names and corresponding ground snow loads as specified in the Supplement. The data for 120 locations in the province of Quebec are currently stored in a table form. When the consultation session starts, the user will be asked to input the city name where the building will be located. If the city name input by the user is not in the data base, then a warning message is given to the user and the user may input another city name closer to the location being considered. When the city name is found in the data base, the system retrieves the corresponding ground snow load and the control is transferred to the rule base for estimating the adjustment factors. Use of a data table for climatic information makes book keeping much easier. Whenever the Code is revised or updated with changes to climatic data, the corresponding changes to the data table can be done without disturbing the rule base. This data table can be expanded to incorporate other design information such as design temperatures and wind pressures and this data table can be shared by more than one rule-base. Significant savings in data base creation by avoiding the duplication of information is an immediate outcome of this approach.

The evaluation of adjustment factors are done by translating the Code specifications into rules. The number of load cases to be considered and the adjustment factors for each case depends on the roof type. The user is given a menu of six different roof types which includes all the roof shapes considered by the Supplement to the Code. Exceptions are the unusual roof geometries and roof areas with obstructions or projections. The user will be asked to provide more input information depending on the roof type. For example, in the case of roof types other than flat roofs, the slope angle would be required to find the slope factor.

About forty rules have been used to encode the information provided in section 4.1.7 of the NBCC. The variable definition provides a 'find' option which eliminates the rules required for prompting the user for input when the value of the variable is unknown. Input validations can also be done as part of the variable definitions. Information required for starting and completing a consultation can be grouped and specified in the initialization and conclusion phase of the rule base. These options and the programming language features significantly reduced the number of rules required in developing the snow load estimation system.

A consultation session with the snow load estimation system is shown in fig.3. The number of load cases and their details are obtained as output based on the location of the building, slope of roof and dimensions of the roof. Arched and curved are one of the most difficult type for which the snow load estimation requires a careful interpretation of the Code. The accumulation factor evaluation and the type of load distribution depends on the rise to span ratio and the height of arch at which the roof slope is 30 degrees (h30). The detailed investigation on the provisions of NBCC 1985 is reported by Kennedy et al (8).

An explanation facility is available to the user for obtaining the intermediate values of the variables used in the process.Reasoning for the rules has not been specified in the current system as the problem is relatively self explanatory. Guru provides about fifty environment variables by which the inference process may be controlled. In the implementation reported here, a forward chaining methodology is used as the system is mostly data driven.

```
       Welcome to the snowload estimation routine
Where is the building located?  Baie-Comeau
Choose the type of roof from:  1 - Flat roof
                               2 - Shed/single sloped roof
                               3 - Gable roof
                               4 - Arch roof
                               5 - Valleys in roof
                               6 - Multi-level roof
Input the type of roof:   4
What is the height of the arch(m)?   4.0
What is the span of the arch(m)?   18.0
What is the edge slope of the arch (in deg.)?   45.0
What is the h30 for this arch?    2.5
Load Case 1:
     Snow load at mid span .....    4.1 kPa
     Snow load along the edges..    2.5 kPa
Load Case 2:
     Snow load (maxm. accumulation)          6.00 kPa
     Snow load along the edge  .....         3.75 kPa
(Refer to fig.H-2 of the NBCC supplement)
*** End of snow load computation ***
```

Fig.3. Example consultation for an arch roof

CONCLUSION

The implementation methodology used in developing a rule-based system for snow load estimation is presented in this paper. It is shown that a data base interface to a rule base can be used in automating Code specifications which require the handling of large amount of data. Knowledge-based systems for building Codes would provide the designer with a consistent interpretation and complete checking of the Code specificationswhich are tedious tasks. The snow load estimation system reported here may be extended and customized to perform the design of roof structures considering the required load cases, load combinations and load distributions. Rule-based systems are suitable for automating Code specifications such as the load estimation, but may not be sufficient to represent all the other information provided by the Code. There are potential problems to rule-based systems when the number of rules increase (10). Considering the limitation of rule-based systems, and if the problem is broken down to small units, then Guru can be effectively used in automating the Code specifications which can be easily maintained.

REFERENCES

1. Fenves,S.J., Maher,M.L., and Sriram,D., "Knowledge-Based Expert Systems in Civil Engineering", Proceedings of the Third Conference on Computing in Civil Engineering, Edited by C.S.Hodge, American Society of Civil Engineers, NY, April 1984.

2. Gero,J.S., and Coyne,R., "The Place of Expert Systems in Architecture", CAD 84, Butterworths, Guildford.

3. Fagan, M.J., "Expert Systems applied to Mechanical Engineering Design - Experience with Bearing Selection and Application Program", Computer Aided Design, Volume 19, No. 7, September 1987.

4. Rosenman,A., and Gero,J.S., "Design Codes as Expert Systems", Computer Aided Design, Volume 17, No. 9, November 1985.

5. Associate Committee on National Building Code, National Building Code of Canada 1985, National Research Council of Canada, Ottowa. Ont.

6. Associate Committee on National Building Code of Canada, The supplement to the National Building Code of Canada 1985, National Research Council of Canada, Ottowa, Ont.

7. Taylor, D.A., "Roof Snow Loads in Canada", Canadian Journal of Civil Engineering, Volume 7, No. 1, March 1980.

8. Kennedy, T.H.R., Kennedy, D.J.L., MacGregor, J.G., and Taylor, D.A., "Snow Loads in the 1985 National Building Code of Canada: Curved Roofs", Canadian Journal of Civil Engineering, Volume 12, No. 3, December 1985.

9. Guru : Reference Manual Volume 1 and 2, Micro Data Base Systems Inc., Lafayette, IN, 1985.

10. Golden, M., Siemens, R.W., and Ferguson, J.C., "What's Wrong with Rules? ", 1986 IEEE Western Conference on Knowledge-Based Engineering and Expert Systems, CA, June 1986.

PC PLUS + LOTUS 123

Siripong Malasri[*], A.M. ASCE

Abstract

A steel beam design program was developed using the LOTUS 123 spreadsheet program and the Personal Consultant Plus, or PC Plus, expert system development package. The expert system was used as a control module and as a design evaluator. The LOTUS 123 spreadsheet was used for a preliminary design and a data base management. Graphics images were created using the TelePaint graphics editor and were integrated into the expert system which makes the system more understandable. An external program written in BASIC was also developed and integrated into the expert system for computing the allowable bending stress for non-compact sections. Programming techniques are discussed.

Introduction

In recent years, there have been several expert system development packages, or shells, available on micro-computers. These shells, however, are not the ultimate tool that can practically solve all classes of problems. Using these shells with other software packages are, therefore, desirable in many applications. This paper describes several software integration techniques of the Personal Consultant Plus, or PC Plus, expert system development package from Texas Instruments (1986) using a steel beam design example.

In a computer-aided steel design process, it is desirable for the program to access a steel section data base. It is not practical to put section properties in a knowledge base as was done in a previous work by Malasri (1986-1987). A data base can, however, be easily and practically handled in the LOTUS 123 spreadsheet (1985) and integrated into an expert system using the EXSYS shell (1985) as shown in another work (Malasri, 1988a). The work described in this paper is somewhat similar to this previous work in main ideas. The PC Plus shell, however, was used instead of the EXSYS shell due to several flexibities it has over the EXSYS shell such as the using of frames and graphics interface.

---
[*] Assistant Professor, Department of Civil & Architectural Engineering, University of Miami, Coral Gables, Florida 33124.

Integrated Steel Beam Design Package

The integrated steel beam design package consists of three modules; the PC Plus expert system module, the LOTUS 123 spreadsheet module, and the BASIC module. The expert system module is the control module that brings up graphics images, executes spreadsheet and BASIC modules, evaluates the trial section in accordance with the AISC specification (1980) provides a list of conclusions to the user, and justifies conclusions by showing rules that were used to obtain conclusions. The spreadsheet module performs a preliminary design, extracts candidate sections from a steel section data base, provides the user a list of trial sections, and transfers information back to the expert system as well as the BASIC module. The BASIC module determines the allowable bending stress for a non-compact section since complex computations are not appropriate in the expert system module, it also transfers the allowable stress back to the expert system. The schematic diagram of the integrated package is shown in Figure 1.

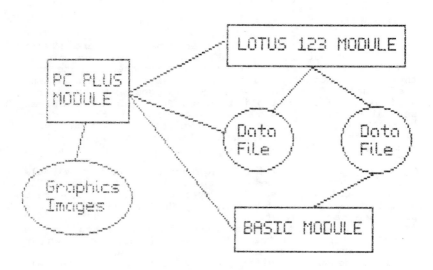

Figure 1. Schematic Diagram of the Integrated Package.

Preliminary Design and Data Base Management

The preliminary design and data base management are performed using the spreadsheet module. Input parameters, i.e., the maximum bending moment and shear, the yield strength of steel, and the unsupported length, can be easily input into the spreadsheet. The preliminary design is done based on the allowable bending stress equaling 0.6 of the yield strength. A -15 % to + 15 % range of the section modulus is used as the search criteria for a data base extraction. Sections that fit in the section-modulus range are extracted and displayed in an output table as shown in Figure 2 where the user can select a trial section. Properties of the selected trial section and other design information are saved into two data files due to different required data file formats; one for the expert system module, another for the BASIC module. The transfer data in the spreadsheet is shown in Figure 3. This spreadsheet module is menu-driven and controlled by macro commands such that the user needs no spreadsheet background. An example of macro commands that automate the data base extraction is shown in Figure 4 with a brief explanation for each command.

	R	S	T	U	V	W	X	..	AA
1		OUTPUT TABLE:							
2									
3	NO	DESIGNATE	d	tw	bf	bf/2tf	d/tw	..	Sx
4	1	"W 36X300"	36.74	0.945	16.655	5	38.9	..	1110
5	2	"W 36X280"	36.52	0.885	16.595	5.3	41.3	..	1030
6	3	"W 36X260"	36.26	0.84	16.55	5.7	43.2	..	953
7	4	"W 36X245"	36.08	0.8	16.51	6.1	45.1	..	895
8	5								
9	6								
10	7								
11	8								
12	9								
13	10								
14	11								
15	12								
16	13								
17	14								
18		App.Sx=		1000					
19									
20	*	SELECT NO.=		3	<----enter your choice here				

Figure 2. Data Base Output Table.

	K	L	M	N	O
21	DATA PASSED TO PC Plus FROM LOTUS 123				
22					
23					
24	Variable		Value		
25	---				PRESS
26	d	(	36.26	)	any key
27	tw	(	0.84	)	to pass data
28	bf	(	16.55	)	to PC Plus
29	bf/2tf	(	5.7	)	
30	d/tw	(	43.2	)	
31	rT	(	4.34	)	
32	d/Af	(	1.52	)	
33	Sx	(	953	)	
34	Fy	(	36	)	
35	Mmax	(	1800	)	
36	Vmax	(	20	)	
37	LU-FT	(	40	)	
38	LU/rT	(	110.59907	)	
39	Section	(	"W 36X260"	)	
40					

Figure 3. Transfer Data in Spreadsheet.

Macro	Explanation
\B	name of macro
/reS4..AA17~	erase range S4..AA17
/reT20~	erase cell T20
{goto}TABLE~	bring up screen with output table
/dqeq~	extract record from data base
{goto}T20~{?}~	wait at cell T20 for user input
{goto}MENU~	bring up main menu
{goto}P43~	highlight cell P43

Figure 4. Automation of Data Base Extraction Process with Macro Commands.

Trial Section Evaluation

The expert system reads the design information from a data file. It first determines local buckling conditions for the flange and the web. It also determines the lateral buckling condition of the member. The system then classifies the section as compact, partially compact, or non-compact. If the section is compact or partially compact, the allowable bending stress is computed in the expert system since the computation is simple. On the other hand, the allowable stress is more convenient to be determined in an external program written in BASIC for a non-compact case. Figure 5 shows additional input values required by the BASIC module.

Is moment at any point within an unbraced length is larger than that at both ends of this length (Y/N)? n

What is the value of the smaller end moment (M1)
--- cw + ? -100

What is the value of the larger end moment (M2)
--- cw + ? 200

Figure 5. Addition Input Required by the BASIC Module for Non-compact Cases.

The moment capacity of the section is then determined and compared with the required bending moment. A similar comparison is done for the shear capacity. A list of conclusions pertaining to the moment and shear is then displayed. Figure 6 shows a sample of a conclusion screen. At a conclusion screen, the user can ask to see rules that were used to obtain a particular conclusion or to compute a value of a parameter. Before ending the consultation, the expert system asks the user if a new trial section is to be investigated. If the user indicates "yes", the system automatically reloads the spreadsheet program where the user can reselect a new section and the evaluation process is repeated until the designer satifies with the selected section.

```
                    STEEL BEAM DESIGNER
:. Conclusions:.........................................
:                                                       :
: Trial section is as follows:  W 36X260                :
: Section condition is as follows:  Non-compact         :
: Moment capacity in k-in is as follows:  20376.64574   :
: Maximum moment in k-in is as follows:  21600          :
: Shear capacity in k is as follows:  438.60096         :
: Maximum shear in k is as follows:  20                 :
: Moment condition is as follows:  Insufficient, select :
:   a larger section                                    :
: Shear condition is as follows:  OK                    :
:                                                       :
:    ** End - RETURN/ENTER to continue                  :
:.......................................................:
```

Figure 6. Sample of Conclusion Screen.

Programming Techniques

a) Frames:

The PC Plus shell allows a knowledge base to be divided into several smaller groups where each group resides in a frame. A frame has its own properties such as goals, initial data, etc. A system may contain only a root frame, or a root frame with several sub-frames. A root frame is always activated first when the system is executed. There are several ways a sub-frame is to be activated. A sub-frame could be activated only once, at least once, atmost once, or it could be specified as an unknown number of occurrences. In the steel beam design package described in this paper, all rules in the expert system are in a sub-frame with the number of occurrences specified as at least once. The root frame does not contain any rule, it contains only goals which are determined in the sub-frame. By doing this, once the evaluation is done and a set of conclusions is made for a given trial section, the user is asked if a new trial section is to be evaluated. The sub-frame can then be re-executed as many times as desired by the user without leaving the expert system. In the previous work (Malasri 1988a), the user has to leave the expert system and starts all over again for each new trial section evaluation. The analogy of using root frame and sub-frames in the PC Plus shell is similar to the use of a main program and sub-routines in FORTRAN.

b) External Program Interface:

Most expert system shells allow external program interface which means that other external programs can be executed from the expert system. This feature is extremely useful since there are several limitations on expert system shells such as a complex procedural computation, a data base management, etc. Several times there exists programs written in other languages, it is therefore desirable to interface an expert system with these existing programs rather than developing a new one. The PC Plus package has two useful functions, i.e., DOS-CALL, and READ-FROM-FILE. The former is used to execute any executable file in DOS, while the latter is utilized to read information back from a data file generated by the external program. Figure 7 shows a rule that executes several external programs, includes the spreadsheet program and reads data back to the expert system. In this application, a utility program is executed before the spreadsheet in order to rename the worksheet file name to AUTO123.WK1 for automatic file loading in spreadsheet. Another utility program is executed after the spreadsheet in order to rename the AUTO123.WK1 back to its original name and to rename the data file from an extension PRN to RD. A PRN extention is provided by the spreadsheet program when the data is saved under PRINT FILE commands, an RD extension is required by the PC Plus expert system. Data to be read into the expert system has to be parenthesized as shown in Figure 3.

IF: PRELIM-DESIGN = "yes"

THEN: DOS-CALL "STEELBM1.EXE" "" AND DOS-CALL "C:\LOTUS\
        123.COM" "123.SET" AND DOS-CALL "STEELBM2.EXE" ""
        AND READ-FROM-FILE "STEELBM" D TW BF BF2TF DTW RT
        DAF SX FY MMAX-K-FT VMAX LU-FT LRT DESIGNATE AND
        ...

Figure 7. External Program Execution from PC Plus.

c) Graphics Interface:

Graphics can be easily integrated into the PC Plus expert system. Graphics images can be created using a graphics editor and brought up during the consultation of the expert system. In this paper, graphics images were created using the TelePaint package (LCS/Telegraphics 1985) for title screen, explanation screen, etc., as shown in Figures 8 and 9. More details of graphics interface of PC Plus can be found in a recent work (Malasri 1988b).

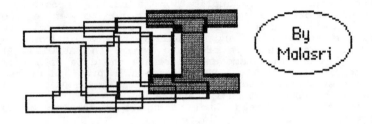

Figure 8. Title Screen.

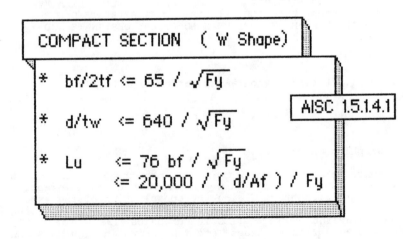

Figure 9. Explanation Screen.

Conclusion

The integration of the PC Plus expert system with the LOTUS 123 spreadsheet program was presented. The PC Plus shell has several other commands for dBASE (Ashton-Tate 1985) data base access as shown in another work (Malasri 1988c), but the method described in this paper is more appropriate and more practical for this application. The external program interface described could work well with other shells. Each software package has its own merit and limitations. Making use of the best features of each package and integrating them together are appropriate and desirable.

References

AISC (1980), *Manual of Steel Construction*, 8th ed.

Ashton-Tate (1985), *dBASE III Plus*.

EXSYS, Inc. (1985), *EXSYS Expert System Development Package*.

LCS/Telegraphics (1985), *TelePaint*.

Lotus Development Corp.(1985), *LOTUS 123*.

Malasri, S. (1986-1987), "Steel Beam Analyzer", *Journal of Structural Engineering Practice*, Marcel Dekker, Inc., Vol. 3, No. 1, pp 55 - 58.

Malasri, S. (1988a), "Expert System - Spreadsheet Integration, to be published, *Journal of Structural Engineering Practice*, Mercel Dekker, Inc., Vol. 4, No. 1.

Malasri, S., and Kengskool, K. (1988b), "Graphics Interface of an Expert System", to be published, *Proceedings of the NCGA'88*, National Computer Graphics Association.

Malasri, S. (1988c), "Data Base Access Using An Expert System Shell", to be published, *Proceedings of the ICES'88 Post-Conference Seminar*.

Texas Instruments Inc.(1986), *Personal Consultant Plus*.

A KNOWLEDGE BASED EXPERT SYSTEM FOR THE SELECTION OF STRUCTURAL SYSTEMS FOR TALL BUILDINGS

P. Jayachandran, A.M. ASCE.,[1] N.Tsapatsaris, A.M. ASCE.[2]

Introduction

With the increasing demand for tall buildings in the US, many new and innovative structural systems have evolved in the past few decades. The structural analysis and design aspects of tall buildings involve experiential knowledge, accumulated in years of practice in structural systems behavior. These systems are created and implemented by a small number of practicing professionals who understand tall building design relatively well.

The majority of software existing today will perform a narrow and routine algorithmic analysis. However, the preliminary analysis and design have not yet been codified. Recognizing the importance of the decisions made in the initial phases of design the validity of such a tool is apparent.

Developments in artificial intelligence have shown that it is possible to simulate the human thought process through the use of "intelligent" solutions strategies. The intellectual effort involved in developing such a program is called Knowledge Engineering. Computer programs which employ knowledge to solve problems that ordinarily require human intelligence are called Knowledge Based Expert Systems.

A Knowledge Based Expert System is presented herein, for the selection of a preliminary structural system for the design of tall buildings. Requesting information from the user on location, height to width ratio, type of occupancy, labor skills, advise on the following aspects is given:

(i) Material of construction.

(ii) Types of structural systems to be used with factors of certainty in descending order of importance.

(iii) Gravity and lateral resisting systems.

The candidate systems are then evaluated by the structural engineer by performing the following [14]:

(i) approximate analysis and design;

(ii) preliminary member sizes selection;

(iii) assessment of drift, stress levels, human comfort, and code compliance of these parameters;

(iv) assessment of steel weight per square foot in both structural steel and reinforced concrete, or mixed/composite systems;

(v) evaluation of systems for fabrication and construction cost optimization;

(vi) evaluation of HVAC and cladding curtain wall systems for cost comparisons

---

1. Associate Professor of Civil Engineering, Worcester Polytechnic Institute, Worcester, MA 01609.

2. Structural Engineer, Milton Chazen Associates, Poughkeepsie, NY 12603.

STRUCTURAL SYSTEMS SELECTION

Structural Systems for Tall Buildings.

The selection of a suitable structural system for a tall building essentially involves three stages of analysis. Design and synthesis with cost comparisons or optimization is made in each [13]:

(i) The initial selection of a structural system with some coordination of functional, mechanical and electrical, requirements for serviceability and people movement; this culminates in studying various types of structural systems and selecting systems such as framed tube, braced frames or shear wall-frames, for further optimization.

(ii) The second stage involves a detailed examination of specific systems to given gravity and lateral loadings and performing an assessment of preliminary member sizes and resulting decision on connections and fabrication cost. The structural engineer is often concerned about weight/square foot of steel for the structural system. This involves an extensive study of substructures, often 4 to 5 floors for gravity and lateral loads. Some parametric examination of girder, column, brace and system geometry is often undertaken to minimize total drift, story drift and human comfort or ductility, in case of wind and earthquake load respectively.

(iii) The third stage involves some extensive final structural analysis and design. Linear dynamic and or inelastic dynamic analysis are employed in 2 and 3 dimensions for lateral time dependent loads. Detailed examinations of connections and fabrication of "structural trees" and their transportation to the site are also made. Prefabrication is always considered. Field welding is minimized as much as possible, to accelerate construction and reduce cost.

The essential structural systems commonly used in most tall buildings can be classified broadly as follows:

1. Moment resisting plane and space frames.

2. Braced frames with concentric and eccentric bracing.

3. Dual systems consisting of moment resisting and braced frames in the interior cores, combining 1 and 2; (e.g., Citi-Corp Center, NYC).

4. Framed-tube systems and Diagonal Trussed-tube systems; (e.g., World Trade Center Towers, NYC, and the John Hancock Tower, Chicago).

5. Partial-tube systems which involve Framed-tubes developed only with exterior flanges of the building framing system; (e.g., 35 story Mercentile Bank Tower, St. Louis, Mo.).

6. Bundled-tube systems which combine several Framed-tubes with belt trusses, located at different levels; (e.g., Sears Tower, Chicago).

7. Composite or mixed structural systems which utilize steel frames with reinforced concrete cores or RC encased structural steel elements such as girders and columns forming composite steel frames with interior shear walls, (e.g., Texas Commerce Court Tower, 75 stories, Houston).

8. Interior braced frames or shear trusses in combination with belt trusses and outrigger trusses which connect them to exterior columns thus enhancing the drift reduction capability of the system; (e.g., First Wisconsin National Bank Tower, Milwaukee).

Figure 1.0 illustrates the classification of various structural systems by Iyengar and Colaco.

Knowledge Based Expert Systems (KBES).

KBES are in many ways different than algorithmic programs. Unlike traditional computer programs, expert systems can be used to solve problems that are unstructured and where no formal procedure exists for finding a solution. The expert system can use its own procedure based on the problem.

Successful expert systems should be a good simulation of the human thought process. Unfortunately the full power of the human brain has not been duplicated. The human mind can use formal reasoning, numerical processing, analogy, intuition, and subjective reasoning in solving problems. With most of the present AI languages one is limited primarily to objective formal reasoning.

Many KBES languages, such as Prolog have codes that are readable and understandable to some degree because they resemble English. This is a valuable aid in the development of expert systems by experts who are not computer specialists. The "transparency" of KBES languages promote easy learning and interpretation.

In KBES the Knowledge-Base and control mechanism are separated. An inference mechanism manipulates the Knowledge Base (KB). This allows for easy expansion and modification of the Knowledge Base without major restructuring, whereas in an algorithmic language modifications are much more difficult. The KB contains two types of knowledge, factual and procedural. A description of the information necessary to solve the problem is part of the factual knowledge.

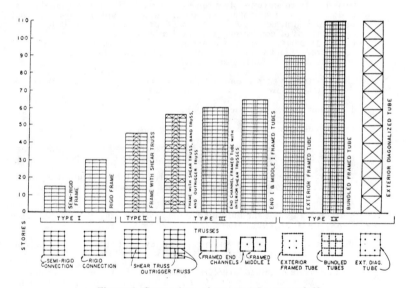

Figure 1. Comparison of structural systems [13]

A complete KBES for the selection of structural system will incorporate other phases of the design such as the computation preliminary sizes and even a detailed analyzer. Such a system is currently under development at WPI and MIT. The modules that will comprise the expert system are shown in figure 2. [3]

In this work the authors were primarily concerned with the *Conceptualizer* which benefits most from the use of a natural language such as Prolog. The *Preliminary Sizer* will compute initial member sizes which will then be analyzed by the *Preliminary Analyzer*. Various stages of the design are guided by the

Fabrication and Erection Reviewer. The *Detailed Analyzer* will contain deeper knowledge, much of which has already been codified in existing stuctural analysis programs.

Modeling the Human Thought Process in the Selection of Initial Structural Systems.

An objective of the research was to identify the criteria with which experts choose structural systems for tall buildings. Essentially, for given architectural and functional requirements a preliminary structural system is selected based on previous experience. This heuristic knowledge can be compiled and stored in a database, but more important than the accumulation of rules and facts was the *modeling of the human thought process.* Thus, the ability to ration and evaluate the expert system database in an intelligent fashion is an important part of the expert system.

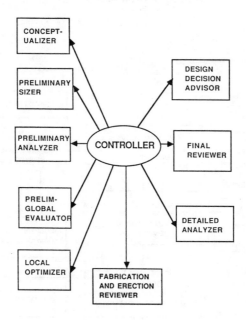

Figure 2. Knowledge Modules [3]

Through interviews, case studies and literature research a broad set of parameters have been identified. The reason for this is that the uniqueness of tall buildings does not allow for isolated design efforts of a structural engineer only. The conceptualization of a structural system requires input from contractors and architects as well as structural engineers. Thus, all areas of expertise are taken into consideration in the selection process.

The most prominent questions tall building design team are: building location, building type, span sizes, floor load, degree of shop fabrication, degree of supervision, availability of labor, quality of the labor, flexibility of the anticipated layout, number of stories, plan aspect ratio, and height aspect ratio.

As for the structural engineer, a systematic methodology of the selection of an initial structural system for a tall building is illustrated in Fig. 3.0. The key steps involved are as follows [16,3]:

> (i) The initial member sizes for gravity loads (dead loads and reduced live loads) are obtained using tributary areas. This holds true for plane frame wall systems but for equivalent tube systems an approximate analysis is made using subassemblies at every 4 to 8 floors. The steel weight is computed with some optimization of column girder stiffness at theses levels.
>
> (ii) Column sizes are determined with approximate wind shears and moments using average column stress levels, reduced to include the effect of gravity and wind moments

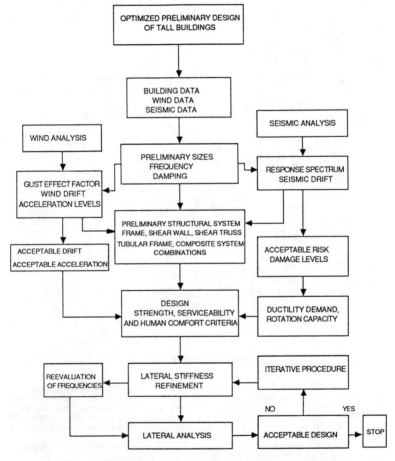

Figure 3.0 The Methodology of Tall Building Design [16,23]

STRUCTURAL SYSTEMS SELECTION 93

(iii) An approximate analysis is made for lateral loadings modeling the building as an equivalent shear flexure cantilever, by using a differential equation solution and determining resultant moments and shears in the wall and frame subsystems acting as flexure and shear beam respectively.

(iv) An approximate story level optimization is made for balanced column-girder stiffness at this time with I_g/L and I_c/L ratios at various story levels. This is to compute steel weights and lateral drift values.

(v) For plane frame type structures, typical one story subassembly, assuming points of inflection at mid-points of columns and girders, is optimized for the story shear at any level and specified drift.

The Knowledge Based environment shown in figure 2. essentially captures all procedures explained in the precedding methodology.
A substantial part of the preliminary structural system selector deals with shallow knowledge that can not be efficiently represented by a procedural language. Turbo Prolog's searching capabilities can model the initial stages of the above described thought process quite effectively.

The Turbo Prolog Programming Language.

The name Prolog comes from the phrase "programming in logic". Prolog is unique in its ability to *infer* (derive by formal reasoning) facts and conclusions from other facts. The user attempts to describe the problem in a logical and coherent manner, presenting facts and knowledge about the problem and specifying a goal. The computer uses this knowledge to achieve the specified goal, defining its own procedure.
Developing programs in Prolog is dramatically different than Fortran or Pascal. Like other implementations of Prolog, Turbo Prolog is an *object-oriented language*. No procedures and essentially no program is used, only data about objects and their relationships are utilized. Emphasis is placed on *symbolic processing*.
While most of the features of the Clocksin and Mellish Prolog are available with Turbo Prolog, the compiled programs execute faster than any other programs on the market for the IBM PC and compatible computers. Functional interface with other languages is available, allowing for procedural support to be added to any Prolog system [27].

Why Turbo Prolog ?

Turbo Prolog lends itself well to the codification of a thought process. The qualities of Turbo Prolog must be carefully implemented so that an accurate simulation can be made. Some of these qualities are its inherent backtracking capabilities and its relentless search for solutions. Prolog will search for a solution until a solution is found or it is told to stop. The human mind works in a similar way. A human expert will think and search his memory until a solution is found or he gives up. Turbo Prolog is quite efficient for symbolic processing and decision making. The same routine can be programmed in Fortran or Pascal. The only difference would be approximately nine times as many lines [27]. This is due to the fact that in the procedural languages, the "backtracking" or search procedure must be programmed as well.
Turbo Prolog does not have the numerical processing capabilities of a procedural language. Thus numerical processing would be limited to boundaries and calculations that have already been made. This is similar to how a human expert thinks. The thought process of a tall building expert does not act as a calculator. Experts remember numbers and can hardly perform sophisticated calculations in their minds. With this philosophy the author believes Prolog can be used to model the human thought process while using other tools to perform algorithmic work. For example at WPI there exist numerous programs which compute wind and seismic loading (Jayachandran) [16]. An attempt to codify these procedures in a natural language would be both wasteful and inefficient. The interfacing capabilities of Turbo Prolog allow for future linking to these programs.

The development of Knowledge Based Expert Systems does not end when the codification is complete. Incorporating the flexibility to accept additional knowledge and valuable criticism resulting from the industry is also an issue. As with many other AI languages Turbo Prolog allows for easy alteration and expansion.

The Program Architecture of the Structural System Selector.

A Turbo Prolog program consists of two or more sections. The main body of the program and the *clauses section* which consist of facts and rules. The clauses section is defined by a *predicates section*. Each relation in each clause must have a corresponding definition in the predicates section. The only exceptions are the built in predicates of Turbo Prolog. The type of object is defined in the *domains* section. This is similar to declaring variables in other languages [27].

The first step in the process is the selection of the construction material. This module will choose a material based on the location of the building. Due to the shallowness of this knowledge it is important to allow for user interface at the material selection stage. If the suggestion is satisfactory the system will then proceed to select a lateral and gravity system. Otherwise the user is given the opportunity implement the material of his choice. Recognizing the importance of a well integrated structure, the lateral and gravity systems are evaluated as a total system to assure their compatibility. A flow chart for the structural system selector is shown in figure 4.

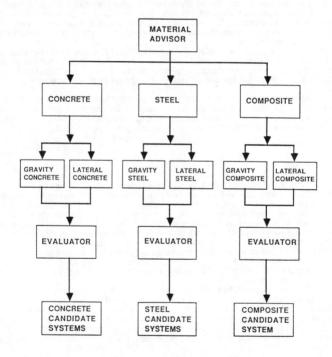

Figure 4. System Flow Chart

… STRUCTURAL SYSTEMS SELECTION 95

Database Acquisition.

Once the most prominent questions of the design team have been identified, one can begin to accumulate the database. Due to the great deal of shallow knowledge involved this process is not a trivial one.

The available knowledge resources must be tapped efficiently to extract the information that is needed. The sources of knowledge should be diverse and accurate. In this work the two major sources for the database have been industry experts and an extensive literature research.

The compilation of the tall building database consisted of accumulating knowledge in the form of facts with respect to the queries generated by the design team. At some point during this process there will exist a great deal of knowledge in the form of rules and facts. It is extremely important to have an efficient *"knowledge transporter"* so as not to lose valuable knowledge in the process. This may take the form of an organized charting and clustering procedure where facts are clustered in similar groups and then charted for ease of codification.

Construction Material Evaluator.

The objective of this module is to provide an intelligent answer as to the construction material to be used based on the location of the building. Thirty cities in the US with significant tall building construction were analyzed.

The evaluation was a combination of two facets. The first being an observation of the number of tall buildings in each city and a breakdown of the construction material used. The second facet gives consideration to cost of construction with respect to the material used. These are total costs based on fabrication, erection and labor costs. A direct comparison of costs is difficult due to the inconsistency in the measurement of work. Therefore by computing a value of cost relative to the national average a comparison of cost between construction materials could be made.

The Conceptualizer then combines both facets of information, thus making an educated selection of a construction material. For example, in New York city almost 100% of the total construction is steel. This complements the fact that concrete construction is 50% more expensive in New York than steel.

The sources of the information for this database were "Mean's Building Construction Cost Data" and "High-Rise Building Data Base" from the ASCE Tall Building Monographs [24,8].

Application of weights to the decision process.

Many decisions are made simultaneously during the design process. Some decisions are more important than others and must be treated accordingly. This prompted the development of a system which will process knowledge and arrive at the best solution.

The decision weights were developed with respect to the decision process of an expert. Each decision made is assigned a weight reflecting its importance. For example, the decision related to the number of stories is assigned a certainty of 85% while for questions such as labor availability a certainty of 20% was used. Structural systems are individually asserted in a dynamic database as a result of the questions asked by the expert system. Each time a system is chosen again its assigned "certainty value" and updated via the procedure shown [28]:

Add Data LC (System, Certain): -
  not(backtrack C(true)),
  Lateral List(System, OldCertain),
  retract(lateraList (System, ___)),
  NewCertain = OldCertain + (100-OldCertain)*(Certain/100),
  asserta (lateralList (System, NewCertain)),!.

In some cases the inherent qualities of Prolog may be disadvantageous. The reason backtracking is temporarily stopped is to allow individual processing of system choices. Backtracking is then re-initiated to continue the procedure for other possible solutions.

The products of the decision procedure are a listing of gravity and lateral systems for which a corresponding certainty value exists. To insure the compatibility of the lateral and gravity systems, a consistency check is performed using the "compatibility" database. This contains information about the compatability of lateral and gravity systems with each other. Possible systems are then listed from best to worst.

The decision weighting process complements backtracking to better simulate the human thought process.

Integration of the Conceptualizer with other modules.

Before the development of a global integration scheme, the integrity of the Conceptualizer was assured. The Conceptualizer consists of three modules within itself (Concrete, Steel, and Composite). Further, the databases are separated from the main module of the program via the standard Prolog *"consult"* predicate. The database consultation only occurs when needed thus not requiring compilation of the whole expert system. *Rather than RAM the only memory constraint for the database is the available disk space.* This allows for an important objective of the expert system to be satisfied: easy expansion and modification of the database.

An important attribute of the tall building expert system environment is an efficient interface between procedures written in different languages. Although Turbo Prolog is an excellent tool for many purposes, there are still many reasons to use other languages. For example, it is easier to perform numeric integration in Fortran; interrupt handling is probably done better in assembly language; and of course if a large program in Pascal already solves some aspect of the problem, this work should not be wasted. Preliminary analysis modules in C can be successful, linked to shallow knowledge modules such as the conceptualizer in Turbo Prolog. Currently, research at WPI has shown that preliminary analysis can be modelled on a micocomputer using C [12].

Turbo Prolog allows interfacing with other languages. The languages supported are Pascal, C, Fortran, and Assembler. The procedures for integrating are simple and can be seen in the Turbo Prolog manual.

Recommendations and Conclusions.

The introduction of Knowledge Based Systems to Civil Engineering has produced an outcrop of expert systems to perform a wide variety of tasks. The realization of the potential in AI has sparked the curiosity of many professionals in academe and industry. Often the interest to apply this new technology does not produce the anticipated results. Unsuccessful attempts to employ Knowledge Based systems are usually a lack of understanding of the qualities and applications of expert system technology.

The ill-structured process of selecting structural systems benefits from a symbolic representation. The manipulation of heuristic knowledge and a search for solutions is a problem that can be handled well with an expert system. It is this aspect of the structural system selection process that must be codified. There would be little use if any, if a language such as Prolog was employed to perform algorithmic work.

The question of whether or not Turbo Prolog is the way to go for the development of a conceptualizer module is a very good one. A definite answer will not exist until other languages have been examined. As a result of this work it is indicative that Prologs's's strong symbolic manipulation, and its inherent search for solution's make it quite attractive and successful in the initial stage of the conceptualizer. Difficulties arise when there is a need for numerical processing to aid the selection process. The last problem is alleviated by the interfacing capabilities of Turbo Prolog.

Discrepancies between output from the Conceptualizer and the actual construction can occur for a variety of reasons. Obviously, problems arise when not enough is known about the specific project. The questions asked by the system are sufficient to make an educated selection of a structural system, but, case-specific parameters may alter the selection. Factors such as politics can not be detected by the expert system as of yet, and will cause problems in many cases.

Future research areas.

An important part of the development of KBES is the knowledge acquisition. More sophisticated methods of compiling and processing this knowledge are necessary to insure adequate knowledge representation.

The conceptualizer is only a small portion of a Knowledge Based Expert System environment for tall building design. Upon completion of other modules a challenging topic will be the integration and process controlling of the total system. Currently, research at WPI and MIT. is examining these and other aspects of KBES.

Example

```
**************************************************************************
*                                                                        *
*                  A KNOWLEDGE BASED EXPERT SYSTEM                       *
*            FOR THE PRELIMINARY SELECTION OF STRUCTURAL SYSTEMS         *
*                                  by                                    *
*                         Nicholas Tsapatsaris                           *
**************************************************************************
PROJECT TITLE : One Mellon Bank Center, Pittsburgh.

In what region of the U.S. will your building be located?
1. Southeast   2. Northeast   3. Midwest   4. Southwest   5. Northwest   6. West
2
Which of these major cities is closest to the proposed site ?
Baltimore, Boston, Buffalo, New York,
Philadelphia, Pittsburgh, or  Washington DC
Pittsburgh
The suggested material for your area is steel. Does the above material suit
your needs (y/n) : y
What is the expected floor span size (feet) : 45
Listed below are various occupancies with their corresponding minimum reduced
live loads in psf.

    1. Assembly halls (60)          9. Scientific laboratories (100)
    2. Offices (80)                10. Hospitals (50)
    3. Card file rooms (125)       11. Penal institutions (40)
    4. Library (150)               12. Mercentile (100)
    5. Schools and Colleges (100)  13. Dormitories (40)
    6. Factories and               14. Hotel or Motel (40)
       Manufacturing (125)         15. Apartments (40)
    7. Foundries (600)             16. Warehousing:
    8. Parking (150)                   light and heavy (200)

Please select an occupancy which best suits the needs of your building: 2
Is the live load above satisfactory?  If not enter a new live load, otherwise
re-enter live load listed : 100
Input the degree of steel shop fabrication that can be performed in your area
     ( 1. very high,   2. high,   3. low) : 1
How would you describe the degree of supervision that can be expected ?
     ( 1. very high,   2. high,   3. low ) : 2
How would you describe the availability of labor in your area ?
     ( 1. very high,   2. high,   3. low ) : 1
How would you describe the quality of steel laborors in your area ?
     ( 1. very high,   2. high,   3. low ) : 2
Input the flexibility of the anticipated layout.
     ( 1. flexible,   2. limited ) : 1
Input number of stories : 54
What is the value of the anticipated plan aspect ratio ( LENGTH/WIDTH )? :1
What is the value of the anticipated height aspect ratio ( HEIGHT/WIDTH )? :4
```

```
***************************************************************
*                                                             *
*              STRUCTURAL SYSTEM SELECTOR OUTPUT              *
*                                                             *
*    Below is a listing of all candidate structural systems.  *
*                                                             *
*              first line = lateral system                    *
*              second line = horizontal system                *
*              third line = degree of certainty (%)           *
*                                                             *
***************************************************************
```

frame_with_shear_truss_band_truss_and_outrigger_truss
vierendeel_girders_with_combined_frame_and_latticed_girders
87

frame_with_belt_truss
vierendeel_girders_with_combined_frame_and_latticed_girders
87

frame_with_shear_truss_band_truss_and_outrigger_truss
welded_I_sections
82

frame_with_belt_truss
welded_I_sections
82

frame_with_shear_truss_band_truss_and_outrigger_truss
latticed_girders
78

frame_with_belt_truss
latticed_girders
78

frame_with_shear_truss_band_truss_and_outrigger_truss
castellated_beams
73

frame_with_shear_truss_band_truss_and_outrigger_truss
stub_girders
73

frame_with_belt_truss
castellated_beams
73

frame_with_belt_truss
stub_girders
73

References
1. Amin, N.R., Taranth, B.S., "Mixed Construction", Chapter SB9, <u>Structural Design of Tall Steel Buildings</u>, Tall Building Monographs, ASCE, Volume, SB, 1979 (Gaylord, C.N., and Watabe, N., Editors)
2. Colaco., J.P., Structural Systems Selection in High-Rise Buildings, <u>Advances in Tall Buildings</u>, Tall Building Monographs, Van Nostrand Reinhold Company, New York, 1986.
3. Connor, J., Jayachandran., P., Sriram, D., "<u>A Knowledge Based Approach for the Preliminary Design of Tall Buildings</u>", Massachusetts Institute of Technology, Worcester Polytechnic Institute, February, 1987.
4. Council on Tall Buildings & Urban Habitat, <u>Structural Design of Tall Concrete and Masonry Buildings</u>, ASCE 1978, pp. 113-115.
5. Council on Tall Building & Urban Habitat, "Concrete Framing Systems", <u>Structural Design of Tall Reinforced Concrete and Masonry Buildings</u>, ASCE, 1978, pp. 58-60.

STRUCTURAL SYSTEMS SELECTION

6. Council on Tall Buildings and Urban Habitat, "Concrete Framing Systems", Structural Design of Tall Concrete and Masonry Buildings, Tall Building Monographs, Volume CB, 1978 pp. 51-91.
7. Council on Tall Buildings and Urban Habitat, "Framing Systems to Resist Horizontal Loads", Tall Building Systems and Concepts, Tall Building Monographs, Volume SC, 1980, pp. 44-51.
8. Council on Tall Buildings and Urban Habitat, "Framing systems to Resist Gravity Loads", Tall Building Systems and Concepts, Tall Building Monographs, Volume SC, 1980, pp. 6-53.
9. Derecho, T. Arnaldo, "Frames and Frame-Shear Wall systems", Response of Multistory Concrete Structures to Lateral Forces, ACI publications SP-36, 1973.
10. Feld, L. S., "Superstructure for 1,350 - ft. World Trade Center", Civil Engineering - ASCE, June 1971.
11. Gasching, JR. Reboh and J. Reiter, "Development of a Knowledge Based System for Water Resources Problems", SRI Project 1619, SRI International, August, 1981.
12. Goldstein, B. "Knowledge Support Modules for the Preliminary Analysis of Tall Buildings" Thesis submitted for MSCE degree, Department of Civil Engineering, Worcester Polytechnic Institute, December, 1987.
13. Iyengar, H.S., "Preliminary Design and Optimization of Steel Building Systems", Proceedings ASCE-IABSE International Conference on Tall Buildings, Vol. 2, Lehigh University, 1972, pp. 185-201.
14. Iyengar, H.S., Composite or Steel-Concrete Construction for Buildings, ASCE, 1977, pp. 61-124.
15. Iyengar, S.H., et. AL., "Elastic Analysis and Design", Structural Design of Tall Steel Buildings, Tall Building Monographs, Chapter SB-2, ASCE, Vol. SB, 1979, (Gaylord, C.N., and Watabe, M. Editors) pp. 31-33.
16. Jayachandran., P., Methodology of Preliminary Design of Tall Buildings, Worcester Polytechnic Institute.
17. Kahn, F.R., "Current Trends in Concrete High Rise Buildings", Tall Buildings, Proceedings of a Symposium at the University of Southampton, Pergamon Press, 1966.
18. Kahn, R.F., Evolution of Structural systems for High Rise Buildings in Both Steel and Concrete, Proceedings ASCE-IABSE International Conference on Tall Buildings, New York, 1973.
19. Kostem, Celal N., "Attributes and Characterstics of Expert Systems", Proceedings ASCE Convention Seattle Washington, April 8-9, 1986.
20. LeFrancois, DR., "Approximate Methods of Analysis and Preliminary Design of Tall Buildings", Thesis submitted for MSCE Degree, Department of Civil Engineering, Worcester Polytechnic Institute, Worcester, MA, July 1981 (Jayachandran.).
21. Ludnegsen J. Philip, Grenney J. William, Drysson Del., Ferrara M. Joseph., "Expert System Tools", Proceedings ASCE Convention Seattle Washington, April 8-9, 1986.
22. Maher, Mary Lou, "Problem Solving Using Expert System Techniques", Proceedings of ASCE Convention, Seattle, Washington, April 8-9, 1986.
23. Modern Steel Construction, Vol. XXIV #3, third quarter, 1984.
24. Means, Building Construction Cost Data, 44Th. Annual Edition, Means 1986, pp. 364-385.
25. Picardi, E. A., "Structural System - Standard Oil of Indiana Building", Journal of the Structural Division, Vol. 99 ST.4, April 1973.
26. Tani, S., "Mixed Construction", Chapter SB-9, Structural Design of Tall Steel Buildings, Tall Building Monographs, ASCE, Volume SB, 1979, (Gaylord, C.N., and Watabe, N. Editors).
27. Townsend, Carl, "Introduction to Turbo Prolog", Sybex, 1987.
28. Tsapatsaris, N., "A Knowledge Based Expert System for the Preliminary Selection of Structural Systems", Thesis submitted for MSCE degree, Department of Civil Engineering, Worcester Polytechnic Institue, July, 1987.

PART II

GEOTECHNICAL AND ENVIRONMENTAL ENGINEERING

GEOTOX-PC: A NEW HAZARDOUS WASTE MANAGEMENT TOOL

George K. Mikroudis[1] and
Hsai-Yang Fang[2]

Abstract

This paper illustrates how knowledge-based expert systems can assist in the identification and remediation of waste disposal sites. Existing methodologies are limited when dealing with incomplete or uncertain data, have difficulties to incorporate the evaluator's risk perceptions into the assessment and may require extensive amounts of data for a reliable application. The paper describes how an expert system termed GEOTOX-PC can remove these limitations by utilizing artificial intelligence techniques, and by relying upon its knowledge base to resolve problems of uncertainties, missing data, or effects of site-specific interactions. GEOTOX-PC is a micro-computer version of a larger system, GEOTOX developed on a mainframe. An evaluation study of GEOTOX-PC, based on six site cases, shows that knowledge-based expert systems introduce a unified appoach that trancends current methods of site assessment.

1 Background

Improper disposal and management of hazardous waste, hazardous substances, and toxic chemicals in numerous and unidentified locations, is one of the most pressing evironmental problems worldwide. The US Environmental Protection Agency (EPA) estimated that approximately 60 million metric tons of hazardous waste are generated annually in the United States at more than 750,000 sites (Silka, 1978). Although not all the disposed wastes are hazardous, EPA estimated that 30,000-50,000 sites have been contaminated and approximately 1,000-2,000 of these sites may threat the environment or pose a hazard to the public. The evaluation of potential hazards at such sites requires knowledge from diverse fields such as geology, hydrology, climatology, chemistry, toxicology, bacteriology, and environmental engineering.

Thus, there is an increasing need to assist the engineer on staying current with technical knowledge from these expanding disciplines. Moreover, for better evaluations, it is desirable to transfer the knowledge and expertise of several specialists for any given problematic site. Existing methodologies vary from qualitative hazard assessment to detailed risk assessment, and numerical modeling of

---

[1]Student Member ASCE, Research Associate, Department of Civil Engineering, Lehigh University, Bethlehem, Pa 18015, USA.

[2]Member ASCE, Professor of Civil Engineering and Director, Geotechnical Engineering Division, Lehigh University, Bethlehem, Pa 18015, USA.

contaminant transport. These methods are limited however when dealing with incomplete or uncertain data, and have difficulties to incorporate the evaluator's risk perceptions into the assessment; in addition, the risk assessment and numerical modeling techniques require extensive amounts of data or prohibitively expensive monitoring and testing for a reliable application. A comprehensive, well established technique that can deal reliably and effectively with all the aspects of hazardous site assessments is still lacking.

These issues inspired the development of a computer-aided consultation program for waste disposal site assessments. An expert system, called GEOTOX, was thus developed at Lehigh University to provide assistance in evaluating a given waste disposal site, and to help the environmental engineer in his decisions and recommendations for remedial actions (Wilson et al., 1984, Mikroudis, 1986a). GEOTOX was developed on a Data General MV-10,000 super-mini computer. The advent of powerful personal computers (PCs) and Artificial Intelligence (AI) software for PCs, enabled porting mainframe software to PCs without severe limitations. GEOTOX-PC is such a micro-computer version of GEOTOX. Using GEOTOX-PC as an example, this paper illustrates how knowledge-based expert systems can be applied to various aspects of site investigation and assessment, and improve the site evaluation process over other existing methodologies. GEOTOX-PC itself is a new tool in decision-making for managing hazardous waste sites.

2 Methods for Waste Disposal Site Assessment

Evaluations of waste disposal sites have been conducted according to a variety of methods, and at various levels of detail (Milner and Roy, 1981). The currently available methods can be classified into three broad categories:
1. *Numerical methods* of groundwater modeling\ contaminant transport
2. Quantitative and formal *risk assessment techniques*
3. Qualitative risk assessment or *empirical models*

All the above methods employ some model of the physical situation at waste disposal sites, in order to make evaluations, interpretations, or predictions. The difficulties and main disadvantages of these methods are briefly described in the following.

Numerical methods The main tranporting agent for land-disposed hazarous wastes is groundwater. Although a great deal is known about the flow of liquids through porous media, the problem of contaminant transport is somewhat more complex. Currently, there are two approaches to the mathematical/ numerical analysis of the problem: (1) Conservative Transport (Anderson, 1983), i.e. contaminant transport without chemical reactions; and (2) Transport with Reactions (Cherry et al., 1983). Using these models, mathematical analysis of the complex processes of contaminant transport is within the current capabilities, but further research and understanding of the phenomena is required. These methods however cannot be used on a stand-alone basis, but need to be supplemented with professional judgment and combined with other factors in order to give a "complete" evaluation requiring extensive data. In addition, since these models are deterministic they cannot explicitly account for uncertainties in the input or in the assumptions underlying the model.

Risk assessment techniques Risk assessment of toxic chemicals is a well-established scientific method which requires comparison of the degree of exposure of individual persons to a toxic chemical with exposure levels known to cause toxic effects. Since exposure is usually difficult to characterize and toxicity information is difficult to extrapolate from animals to humans, risk assessments for toxic chemicals are generally difficult to apply and yield uncertain results (Nisbet, 1984). The main disadvantage of these methods is that they require prohibitively expensive monitoring and toxicity testing data. An additional limitation of most of these techniques which are based on a decision-tree is that they are not flexible and are difficult to modify. There is a predetermined series of questions to be asked which prevents the user from volunteering information, using additional data not anticipated by the model, responding with an answer such as "unknown", or restructuring the hierarchy of the decision-tree without affecting the logic of the entire model. These disadvantages become quite limiting when dealing with incomplete data and with changing and expanding knowledge as often happens during hazardous waste investigations.

Qualitative methods For preliminary assessment of sites with limited field data, qualitative hazard assessment models have been used extensively to yield a relative ranking of sites and to establish priorities for remedial action. The assessment is essentially based upon information related to four primary characteristics as follows (Berger, 1984):
   a) Targets or Receptors (humans or the environment that may be exposed to hazards from the site);
   b) Routes or Pathways (by which hazardous substances may migrate from the site);
   c) Contaminant or Waste Characteristics; and
   d) Waste Management and Control Practices.
The labels may differ depending upon the particular risk assessment model, but the basic areas of concern remain the same. Some of these qualitative site evaluation systems are the LeGrand method (LeGrand, 1980), the Surface Impoundment Assessment model (Silka, 1978), the Rating Methodology Model (JRB Associates, 1980), and the Hazard Ranking System (MITRE Corporation, 1984). The main inadequacy of these models is that their assessments can rarely be quantitatively reliable unless a very detailed monitoring program is carried out. In most cases, they yield a ranking of risk on a qualitative or semi-quantitative scale (e.g. from "low" to "high" or on a 0-10 scale). These methods can only be reliable and useful tools in decision-making for hazardous waste issues, if their practicioners have sufficient knowledge and experience to make the reasonable scientific judgments involved.

This need for expert assistance in decision-making at hazardous waste sites inspired the development of GEOTOX-PC, and the application of knowledge-based expert systems which will be described in the following. The classification approach of GEOTOX-PC is not as rigid as that of the empirical models. The inability of these models to account for parameter changes and interactions of site specific situations is one of their main weaknesses. Moreover, in these models uncertainty in data is not considered explicitly and is difficult to estimate the confidence and the validity of the results. GEOTOX-PC is designed to overcome these limitations by utilizing AI techniques and by using expert knowledge to resolve problems of uncertainty, missing data, or effects of important site-specific interactions. The remainder of the paper presents GEOTOX-PC, its implementation, and how such an expert system is used to improve the evaluation of hazardous waste sites.

3 GEOTOX-PC: A KBES for Evaluating Hazardous Waste Sites

GEOTOX is a knowledge-based expert system primarily designed for hazardous waste site evaluations (Mikroudis, 1986a, Pamucku et al., 1987). GEOTOX was originally developed on a Data General MV-10,000 super-mini computer. A new version of GEOTOX, termed GEOTOX-PC, has been developed for personal computers, without significant limitations. The knowledge-base and inference capabilities of GEOTOX-PC are essentially the same with the mainframe version. GEOTOX-PC is only limited because of the memory limitations of personal computers.

GEOTOX-PC was developed on an IBM AT -- a 286 machine. The new 386 machines and new operating systems (such as OS/2) that take advantage of the extended memory and multitasking capabilities of the micro-processor will enable a full PC implementation of GEOTOX without any restrictions. The current version of GEOTOX-PC is running under DOS and requires at least 640K of memory. It is developed using ADA Prolog, a standard Clocksin-Mellish Prolog (Clocksin and Mellish, 1981). The main problem of this implementation is that because of the 640K barrier of DOS as well as certain file limitations, the knowledge-base cannot be updated in run-time for any realistic application. Thus, this capability of the GEOTOX mainframe-based system was not ported on the PC. Also the shell capabilities of the mainframe system that allow GEOTOX to provide a software development environment for environmental geotechnology applications, are not present in GEOTOX-PC.

GEOTOX-PC is intended to assist in preliminary investigations, although its knowledge base can be expanded to accommodate detailed investigations and field work. It can be used also for multiple site comparisons, prioritizing and ranking. Besides evaluation of existing sites, GEOTOX-PC can be applied to assess potential sites and assist in the evaluation of the site selection process for new facilities. The various processes supported by GEOTOX-PC are summarized as follows:

Interpretation   Assessment of existing hazardous waste sites
                    Evaluation of potential waste disposal sites

Classification   Ranking of existing sites
                    Screening of potential sites

Diagnosis   Contamination problems at hazardous waste sites
                    Selection of remedial alternatives

A GEOTOX-PC consultation mainly assists the user in an interpretation task, by organizing the data, classifying the observations, and inferring possible consequences for the site existing conditions. The inference capabilities of the system allow GEOTOX-PC to be used in diagnostic tasks as well. The system will find what are the potential problems at the site, the possibility for contamination, the type of contamination, and the seriousness of the situation. Finally, by applying GEOTOX-PC, and by varying the input data, the evaluator can study different alternatives for remedial action depending on the site conditions.

3.1 Conceptual Framework of a KBES and GEOTOX-PC

A conceptual framework, which provided a unified approach to the problem has been presented (Wilson et al., 1984, Mikroudis, 1986a). The critical elements of this framework, shown in Figure 1, are:

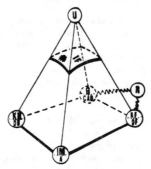

Figure 1: The Wilson-Fang Model of GEOTOX

- The User (U) accesses and controls the flow of information in the system.
- The Knowledge Base (KB) for rules and application knowledge.
- The Data Bases (DB) related to the domain (e.g. TOXLINE, HAZARDLINE, CHEMLINE, CESARS, OHMTADS, TDB, RTECS) (Arthur, 1984).
- The Inference Mechanism (IM) for knowledge processing and modification.
- The database for both user known facts (KF), and deduced facts (DF) from the inference procedure.
- Algorithmic Structures and analysis programs (A), to analyze contaminant transport, ground water flow, etc.
- A 3-dimensional Graphics system (G), plus Computer Aided Design (CAD) package which facilitates geometrical design of containment alternatives (covers, liners, drainage systems, pumping wells), in three dimensions.
- Remote Sensing and robotics subsystems (R) for data acquisition and/or processing in hazardous environments.
- The user Interface (IF) which coordinates and communicates the operational processes and explanations to the user.

The value of this model is that it depicts clearly the general software subsystems (expert system, graphics, and remote sensing) and the parallel processing of information, and aids the visualization of their interactions within a computer-integrated engineering system. It also provides a unified approach that accommodates interpretation, diagnosis, planning, design, and monitoring of hazardous waste sites.

The current GEOTOX-PC prototype, built upon the Wilson-Fang framework, does not include hardware for the remote sensing of data, and has somewhat limited graphic capabilities. The analysis programs in GEOTOX-PC include some empirical formulas and simplified equations of groundwater flow applicable to special cases (e.g. uniform, sandy aquifer). They can be extended, however, in complexity and sophistication from this level up to the use of existing finite element programs.

3.2 Knowledge Base Architecture

The objective of an assessment of a waste disposal site is to describe both what exists and what may happen at the site. A comprehensive evaluation includes a thorough and detailed description of site-related data, waste characteristics and regional characteristics on one hand, and a list of potential hazards or possible problematic situations on the other. The computer representation of the problem, therefore, involves two tasks. First, the site-related data which describe the existing conditions must be organized, and then the expert knowledge which guides the solution must be represented.

In this section we discuss the knowledge representation in GEOTOX-PC and the overall structure of the site evaluation problem. The knowledge representation scheme was determined primarily by the problem characteristics. GEOTOX-PC mainly deals with an interpretative rather than a diagnostic problem. The system, as mentioned, may perform some diagnostic tasks at a time (e.g. decide what contamination problem exists at the site), but in its main mode of consultation it attempts to integrate the observations provided by the user into a concise description of the site and the possible environmental hazards. This interpretation task is more naturally accomplished with a forward type of reasoning, where the observed evidence activates hypotheses, and generates conclusions. The relationships between the data and site characteristics as well as between site characteristics and site parameters or general concepts needed for the evaluation can be conveniently described in terms of relations in an associative network. In a diagnostic task, however, a backward type of reasoning is more appropriate, where hypotheses are tested until one or more are found which satisfy the existing conditions. This is more easily described in a production rule formalism. Finally, when the task is to describe situations in a comprehensive manner, as it happens when GEOTOX-PC is asked for conclusions, all relevant data, reasoning, and results must be brought together for a sufficient representation. This leads to the use of the frame formalism as means for representing the problem. Hence three different methods of knowledge representation (associative net, production rules, frames) each one with its own merits can be applied to different aspects of the site assessment problem.

In order to respond successfully to the representation requirements of the problem, GEOTOX-PC uses a combination of the above methods, built around an associative network, as shown in Figure 2. Inherited values by means of the associative network constitute the basic mode of inference for GEOTOX-PC. This hybrid system of an associative net and production rules is implemented using Logic Programming. The characteristics of these representation methods and their suitability are discussed in the following.

Associative Network The associative network provides the underlying structure to the overall knowledge representation scheme in GEOTOX-PC. This network defines all the associations between data and site parameters as conceived by the domain expert(s). Every site characteristic is represented by a *node* in the network, and this node is associated with others by means of labelled *arcs* (*links*) representing

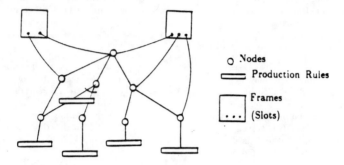

Figure 2: Simplified View of the Knowledge Representation Scheme in GEOTOX-PC

various types of relationships between site characteristics. Although from a conceptual point of view the associative links between the nodes may represent a variety of relations (cause-effect, taxonomies, definitions) when used for inference they can only be one of two types: conjunctive or disjunctive. Disjunctive nodes can propagate their values independently of others, but conjunctive nodes cannot. Therefore, the associative network is used in a conceptual sense to define what is going to affect what. "By how much" is determined by production rules or, in the absence of rules, by the inference mechanism.

Production Rules Production rules are used in GEOTOX-PC to test conditions. They are used at conjunctive nodes of the network as well as at the leaf nodes to determine what value should be propagated to the network according to the existing conditions at the node. They represent the condition-action judgmental statements of the expert and express the estimated level of hazard by a pair of values: a hazard value (on a 0-10 scale) and a confidence level (on a 0-1 scale). The hazard values (defined by the expert) represent indices for the severity of the situation at the site. The confidence level represents the confidence of the expert to the rule and, thus, it is analogous to a weight related to the hazard index. The assignment of the pair of hazard-confidence values to a node according to the applicable production rule represents the use of expert rules of thumb to decide on the effects of given conditions. If such rules are absent then the more generic relationships represented in the associative network are used to infer the pair of values at the current node.

Frames The overall representation scheme, consisting of an associative network with production rules attached to conjunction nodes and to the leaf nodes, is completed with frames attached to various nodes of the network, as shown in Figure 2. Frames, which are a collection of associative net nodes and slots, are used to represent the final conclusions of the expert or to describe different situations or possible scenarios of contamination at the site. Their slots are mostly filled with values from the higher nodes in the associative network. The slots that are activated provide the frame description of a particular aspect of the evaluation.

HAZARDOUS WASTE MANAGEMENT TOOL 109

3.3 Reasoning and uncertainty

When evaluating a site, GEOTOX-PC is trying to establish a level of potential hazard related to the site. The ultimate question is *"How good, or how bad is the site?"*. Usually, the approach taken in such cases is to describe the situation by using an index as a measure of the degree of hazard involved. Such an index is estimated as an aggregate of several other subindices describing other important aspects of the situation. In GEOTOX-PC, a hazard index H is used on a scale from 0 to 10, where higher H indicates higher potential hazard from the site. To a certain extent every site characteristic affects H and this is indicated by a subindex h_i associated with the characteristic. These indices are defined by the expert during the development of GEOTOX-PC and are contained in the knowledge base. In the following, it is explained how they are used and combined together under uncertainty in order to derive the overall level of hazard at any given site.

Inference Mechanism GEOTOX-PC uses the hazard value h and confidence level c as provided by the production rules and propagates them to all the associated nodes following the links in the associative network. Thus, the system asks data corresponding to the leaf nodes of the network and then finds the applicable rule for the given data. The rule fires and a hazard value - confidence level $(h$-$c)$ pair is assigned to the corresponding leaf node of the associative network. The leaf node is connected to one or more parent nodes in the network, which inherit the h-c pair. If the son-parent link is a disjunctive one and no other h-c pair exists in the parent node the new value is simply also assigned to the parent. For a conjunctive node, the applicable production rules are checked and if a rule is find that can fire, a new h-c pair is assigned to the parent node according to what the rule says. In case that a h-c pair already exists at the parent node, say h_o-c_o, it is updated and a new pair, h_n-c_n is found by applying the inference rule:

$$s = N(c), \quad s_o = N(c_o) \tag{1a}$$
$$1/s_n^2 = 1/s^2 + 1/s_o^2 \tag{1b}$$
$$h_n = (h s_o^2 + h_o s^2)/(s^2 + s_o^2) \tag{1c}$$
$$c_n = N^{-1}(s_n) \tag{1d}$$

where, s is the standard deviation of a normal distribution N with a probability of c between h-0.5 and $h+0.5$

Once the new h-c values are found at the parent nodes, the propagate-update process continues upward in the network until all the associated nodes are updated (forward-chaining). For a single piece of evidence each node is updated only once during this process.

The reason for selecting the above equation is that it satisfies commutativity, it has the property of statistically minimizing the error in h and it increases the confidence in c with every successive application (Gelb, 1974). Equations (1b) and (1c) are found by the use of Bayes' theorem. Figures 3a and 3b show the h-c pair at typical nodes i and j of the GEOTOX-PC associative network. Figure 3c illustrates the effect of applying the inference rule for a given h-c pair. It shows that the updated h is between the old and new values and closer to the value provided with more confidence. In the same time the updated confidence is greater than both the

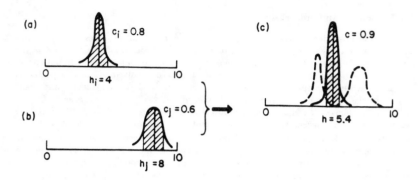

Figure 3: Application of the Inference rule in GEOTOX-PC

old and new values as indicated by the decreased standard deviation of the normal curve. In the extreme case where one of the values is given with zero confidence the effect of the rule is that both h and c remain unchanged.

Flow of Control When GEOTOX-PC is given some data it executes the basic control loop as follows:
  a) update the network
  b) check priorities
  c) ask questions
  d) get new data or satisfy a user request
  e) go to (a)

When the system sets priorities it tries to determine which is the most important question to ask next. The intent is to increase the overall confidence to the results. This happens whenever GEOTOX-PC takes the initiative to ask questions. At any point during the consultation, however, the user can volunteer information (activate pre-established nodes), and thus provide the data he thinks are most important first. Of course, other schemes for asking questions can be followed, the simplest one being just to follow in sequence the initial priority list provided by the expert. In any case, the final result does not depend on the sequence of rules since the basic GEOTOX-PC inference rule satisfies commutativity. The level of confidence in the consultation depends only on the amount and quality of the data provided by the user. Thus, the strategy followed by GEOTOX-PC for guiding the consultation is most effective in the case where the user follows an exploratory process for the investigation, looking for more data and providing more observations until he reaches a satisfactory level of confidence.

3.4 The user-system interface

GEOTOX-PC provides facilities which allow the user to examine the contents of the knowledge base as well as of the current state of the consultation, and to ask how and why questions about the results. Moreover GEOTOX-PC allows the user to volunteer information, change answers to previous questions, deactivate parts of the inference network, and include his own parameters in the assessment. The why-how requests are handled in GEOTOX-PC by tracing up and down the associative network, and by displaying the applicable rules, if any. For volunteering information and changing answers to previous questions the user has to utilize keyword names that are recognized by the system. At the present stage, in such cases, GEOTOX-PC is just repeating the updating of the network from the beginning of the consultation, since this is a simple process and it does not require significant time (except for very lengthy consultations).

User modifications to the knowledge base A unique aspect of the GEOTOX mainframe system interface is the increased flexibility it provides the user by allowing him to include his own parameters in the assessment, or to deactivate certain parts of the associative network. Because of PC memory limitations and limitations of the DOS operating system, this capability was not directly implemented in GEOTOX-PC. In order for a user to make changes in the knowledge-base, he would have to edit the file that contains the GEOTOX-PC knowledge-base and then include it in his consultation. The new 386 machines and more advanced operating systems, such as OS/2, will remove these limitations. A new version of GEOTOX-PC is currently under development for the 386 machines. The new version will allow the user to modify the knowledge-base in run-time.

4 Evaluation and Verification of the KBES

Expert system evaluations are performed during all the stages of their development, but a general, broadly accepted methodology of how to perform such evaluations is lacking. The methodology followed in the case of GEOTOX-PC and the results of the evaluation are presented in the following.

Methodology The main-frame GEOTOX has been in use since early 1986, and has been extensively tested from its developpers, on-site users at Lehigh University, and outside users from environmental engineering firms. Porting the software to a PC required some duplication of this testing for GEOTOX-PC this time. GEOTOX-PC was tested in three different ways, each one concentrating on different aspects of the program:
1. *Domain expert* (professor H.Y. Fang).
2. *Existing methodologies* (LeGrand, SIA, RMM, HRS).
3. *Sensitivity studies.*

The evaluation procedure followed a sequence of steps. Initially, a number of test sites were selected according to criteria set by Fang for the purposes of the evaluation. A number of site factors were also identified to form the basis of subsequent comparisons. Then, for each site, a site documentation form was completed describing the observed characteristics of the site known from the site records. GEOTOX-PC evaluated these data and derived its conclusions about the site expressed on a 0-10 hazard index scale. In the same time, the expert evaluated the sites using the site documentation forms, and gave his conclusions expressed on the same 0-10 scale. Finally, the sites were evaluated according to other methodologies using the same data, where applicable. The six sites were selected from six states and for different reasons shown in Table 1 below.

Table 1: Case-studies for Testing GEOTOX-PC

#	NAME	STATE	REASON FOR INCLUSION
1	Middletown, Road Dump	Maryland	Potential Contamination of Groundwater & Surface Water
2	Voortman Farm	Pennsylvania	Sinkhole
3	Haystack Facility	Oklahoma	Proposed Facility
4	Matthews Electroplating	Virginia	Contamination of Groundwater
5	West Virginia Ordnance	West Virginia	Contamination of Surface water
6	Dover Air Force	Delaware	Potential Contamination of Groundwater

4.1 Discussion of results

Table 2 presents the results of the six case-studies. Figure 4 compares GEOTOX-PC to the results of the expert (HYF) and the four empirical methods (LeGrand, SIA, RMM, HRS). Figure 5 compares the results of the expert to those of the empirical methods. A discussion of these results follows.

Table 2: Results of the Case-Studies

METHOD \ CASE#	1	2	3	4	5	6	maxD %
HYF	6.1	5.9	3.8	6.3	5.2	6.1	
GEOTOX w/ HYF	6.2	5.9	3.5	6.4	5.4	6.0	-3
(confidence)	(0.38)	(0.40)	(0.41)	(0.42)	(0.40)	(0.36)	
LeGrand (HEL)	6.8	8.4	2.9	7.3	6.1	5.5	
GEOTOX w/ HEL	7.5	7.9	3.7	7.3	6.8	5.6	8
(confidence)	(0.41)	(0.45)	(0.31)	(0.40)	(0.42)	(0.34)	
SIA	7.9	9.0	4.5	7.6	8.6	7.2	
GEOTOX w/ SIA	7.1	8.2	4.8	6.8	7.7	7.4	-9
(confidence)	(0.28)	(0.28)	(0.28)	(0.29)	(0.28)	(0.28)	
RMM	6.2	6.8	4.4	6.8	8.8	5.9	
GEOTOX w/ RMM	6.2	6.7	4.7	6.7	6.6	6.0	3
(confidence)	(0.74)	(0.74)	(0.74)	(0.72)	(0.74)	(0.69)	
HRS	6.5	6.8	3.8	6.8	7.1	7.1	
GEOTOX w/ HRS	6.6	6.5	4.5	7.0	7.0	6.7	7
(confidence)	(0.66)	(0.67)	(0.61)	(0.71)	(0.72)	(0.70)	
maxD w/ HYF %	1	0	-3	1	2	-1	-3
maxD w/ Other	-8	-8	8	-8	-9	-4	-9

GEOTOX-PC vs. the expert Table 2 indicates that the maximum difference between the GEOTOX-PC score and the corresponding value of the domain expert for these six cases is 0.3, which is 3% of the 0-10 scale. This score corresponds to the overall conclusions about each site as they are summarized in the overal hazard index *H*. Although this was considered a good agreement, a detailed inspection of the sub-conclusions for each case-study (reported by Mikroudis (Mikroudis, 1986b)) revealed a maximum difference in the sub-conclusions of 1.1, i.e. 11% of the 0-10 scale.

Further, the domain expert identified critical situations and potential problems at certain sites that the program was not able to pick up. For example, in case-study #2, HYF mentioned the possibility of liquid pollution creating new sinkholes. GEOTOX-PC could not generate such conclusions because the necessary knowledge was missing from its knowledge base.

The differences between GEOTOX-PC and HYF for sub-scores was not very discouraging since such differences can be partially attributed to the scoring mechanism of GEOTOX-PC. Sub-conclusions involve far less parameters than the overall conclusions and the GEOTOX-PC assessment improves as the number of parameter increases (as shown by Equation 1). The fact that GEOTOX-PC was not able to identify all the details of HYF's assessment reveals however that the program is not an "expert" yet. This is not surpising because much of the effort in the program's development was put in the system design and programming rather than in detailed knowledge acquisition. The overall performance of the system and the level of agreement between its results and those of the expert indicate that GEOTOX-PC can be further improved by incorporating additional rules or revising existing rules in the knowledge base. The design of the system, however, should not need major changes to improve the performance.

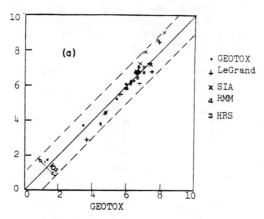

Figure 4: GEOTOX-PC vs. Others

Finally, an interesting result is revealed by comparing Figures 4 and 5. The large deviations from the 45°-line shown in Figure 5 compared to the small deviations of Figure 4 may seem as a surprise. Although GEOTOX-PC-HYF and GEOTOX-PC- other methods agree quite well, HYF does not agree well with other systems. This is evidently attributed to the fact that GEOTOX-PC was compared to other methods by utilizing the exact same data and activating from its knowledge base the same factors that were used by each system being compared. Fang, on the other hand, did not adjust the kinds of factors he was using according to each method.

This result indicates that the scoring mechanism, or the mathematical formulas used to combine the various site factors are less important than the kinds of parameters involved in the assessment. Two evaluation methods using the same parameters and the same data should yield similar results. Also, by inspecting Figure 4 it can be seen that the greater the number of parameters used the smaller the differences are between GEOTOX-PC and the system being compared. These observations also reinforce the fact that the problem-solving power of a knowledge-based expert system, and thus GEOTOX-PC, comes from the *knowledge* it possesses and not just from the formalisms and inference schemes it employs (Feigenbaum, 1977).

GEOTOX-PC *vs. other methods* The rightmost column of Table 2 shows that GEOTOX-PC is in good agreement with the different empirical site evaluation methods, the maximum difference being from 3% to 9% of the 0-10 scale. It should be emphasized again that these methods are quite different to each other considering different factors for the site assessment and using different scoring mechanisms to calculate the overall hazard. Since GEOTOX-PC includes in its knowledge base all the factors that may be used by other systems as well as by Fang, and since the GEOTOX-PC system has the flexibility to use whatever factors are needed in each case, it was possible to compare GEOTOX-PC to each system by activating from the knowledge base only the factors that each given system was using. The close agreements shown in Figure 4 reflect the capability of the GEOTOX-PC system to synthesize many diverse factors, emphasizing the important aspects of the assessment and resolving differences in input data or kinds of factors employed by different evaluation methodologies.

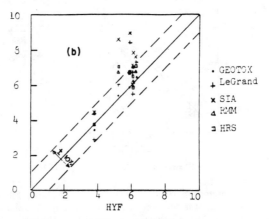

Figure 5: HYF vs. Other Methods

Considering the many sources of disagreement in assessing individual factors for each method (coming from different experts) and the innate uncertainties involved, the capability of GEOTOX-PC to resolve the differences among the different methods and essentially arrive at the same overall conclusions is very important. In a sense the empirical systems can become degenerate cases of a GEOTOX-PC assessment. This result indicates that GEOTOX-PC and KBES, in general, introduce a new unified approach that transcends currently available methods for the identification/ remediation of hazardous waste sites.

5 Concluding Remarks

This paper presented how knowledge-based expert systems can be effectively utilized for the identification and remediation of hazardous waste sites. As an example, GEOTOX-PC, one such system was described and its various subsystems (knowledge-base, inference engine, interface) were explicated. GEOTOX-PC was applied to six case-studies of waste disposal sites and its performance was compared against that of the domain expert who contributed in the system's knowledge base, and against the evaluations of four other site assessment methodologies. Given the uncertainties and variabilities in hazardous waste site assessments, this study showed that knowledge-based expert systems and GEOTOX-PC introduce a new approach that can effectively lead to more objective, repeatable, uniform, detailed and understandable assessments than those obtained from currently available methods. The knowlege base and the user-system interface of GEOTOX-PC make it a versatile and powerful tool for decision-making in the identification and remediation of hazardous waste sites. Practical benefits of using such a KBES include rapid inspection, better knowledge of the physical situation, and improved assessment and remedial decision-making.

5.1 Further research

The integration of computer graphics would serve an important aspect of a KBES such as GEOTOX-PC since it serves as a significant supplement to both the human interface and the inference mechanism within the expert system. It would allow the evaluator to visualize conditions and alternatives, and permits him to deal more easily with the impreciseness and uncertainties of the toxic problem. For example, 3-dimensional color computer graphics could clearly display possible contamination patterns in space and time. Also, associative relationships, such as those shown in Figure 2 could be displayed graphically to enhance the understanding of the inference process in GEOTOX-PC. As part of the current prototype, the authors have developed an interface between GEOTOX-PC and the Graphical Kernel System (GKS) in an initial attempt to add computer graphics to the system.

In addition to expansions of the interface, future efforts in the GEOTOX-PC system or any other KBES for hazardous waste site investigations, should address the possibility of utilizing existing data bases such as TOXLINE, HAZARDLINE, CESARS, RTECS and OHMTADS. This will improve the versatility of the system and will increase its effectiveness especially in the problem of waste characterization. The correct identification of the chemicals and their characteristics is perhaps the most difficult step in the whole assessment. Currently, GEOTOX-PC uses the expert knowledge from its knowledge base as well as suggested EPA guidelines for priority chemicals. The direct access to chemical/toxicological data bases and the development of knowledge base rules to utilize such information will increase the diagnostic power and the reliability of the system.

Also, since GEOTOX-PC was developed based on the contribution of a single expert, it is desirable to expand the system's knowledge base to incorporate the opinion of other experts. The additional contribution from such experts, or a panel of experts, should increase the credibility of a KBES' advice and the overall acceptability of the system. Possible differences in opinion between these experts however, will have to be resolved before they can be incorporated directly in the system's knowledge base. Another possibility is to have different experts contribute to different aspects of the knowledge base such as waste characteristics or geologic characteristics. On the other hand, combining separate knowledge bases from different experts is beyond the current capabilities of KBES and GEOTOX-PC. The use of multiple knowledge bases is a subject of current AI research which can be directly applied to the problem of identification and remediation of hazardous waste sites (Fang and Mikroudis, 1987, Mikroudis and Fang, 1987).

Finally, further research is needed on existing methods of knowledge representation. Associative networks, rules, logic and other AI techniques cannot capture all the aspects of human reasoning or cognitive processes. There is more to intelligence than calculative rationality and knowledge-based systems need to improve in that respect. The understanding of deeper reasoning and the search for methods to represent the knowledge and cognitive processes involved, are beyond the scope of the GEOTOX-PC project, but promising, important areas of future research.

References

Anderson M.P. Movement of Contaminants in Groundwater: Groundwater Transport - Advection and Dispersion. In *Groundwater Contamination in the U.S..* , 1983.

Arthur J. Safety and Health Information for Use in Responding to Hazardous Waste Emergencies. In *National Conference on Management of Uncontrolled and Hazardous Waste Sites.* U.S. EPA, 1984.

Berger I.S. Determination of Risk for Uncontrolled Hazardous Waste Sites. In *National Conference on Management of Uncontrolled and Hazardous Waste Sites.* U.S. EPA, 1984.

Cherry J.A., Gilham R.W. and Barker J.F. Contaminants in Groundwater: Chemical Processes. In *Groundwater Contamination in the U.S..* , 1983.

Clocksin W.F., and Mellish C.S. *Programming in Prolog.* Berlin, Heidelberg, New York: Springer-Verlag, 1981.

Fang, H. Y., and Mikroudis, G. K. *Conceptual Models of Multi-Domain Expert Systems* (Tech. Rep. CS-87-07). Envirotronics Corporation International, June 1987.

Feigenbaum E.A. The Art of Artificial Intelligence: Themes and Case Studies of Knowledge Engineering. *IJCAI*, 1977, 5, 1014-1029.

Gelb A. (editor). *Applied Optimal Estimation.* Gabridge, Massachusetts: MIT Press, 1974.

JRB Associates, Inc. *Methodology for Rating the Risk Potential of Hazardous Waste Disposal Sites* (Tech. Rep.). U.S. EPA, May 1980.

LeGrand H.E. *A Standardized System for Evaluating Waste Disposal Sites.* National Water Well Association, 1980.

Mikroudis G.K. GEOTOX: *A Knowledge-Based Surrogate Consultant for Evaluating Waste Disposal Sites.* Doctoral dissertation, Lehigh University, December 1986.

Mikroudis, G. K. and Fang, H. Y. *Multi-Domain Knowledge-Based Expert Systems and Beyond* (Tech. Rep. CS-87-12). Envirotronics Corporation International, September 1987.

Milner P.M. and Roy S.P. *Ground Water Quality Methods - A Summary* (Tech. Rep.). Commonwealth of Massachusetts, Department of Environmental Engineering, 1981.

MITRE Corporation. *Uncontrolled Hazardous Waste Site Ranking System. A User's Manual.* US Environmental Protection Agency, 1984. (Originally published in the July 16, 1982, Federal Register, Vol 47, No 137, pp31219-31243).

Nisbet I.C.T. Uses and Limitations of Risk Assessments in Decision-Making on Hazardous Waste Sites. In *National Conference on Management of Uncontrolled and Hazardous Waste Sites.* U.S. EPA, 1984.

Pamucku, S., Mikroudis, G. K., and Fang, H. Y. GEOTOX- A New Knowledge/Data Base Management System for Controlling Solid Wastes. In *Environment'87, International Symposium on Environmental Management.* Bogazici University,Istanbul, Turkey: General Directorate of Environment - Pollution Control Research Group, 1987.

Silka L.R. *A Manual for Evaluating the Contamination Potential of Surface Impoundments* (Tech. Rep. EPA 570/9-78-003). U.S. EPA, June 1978.

Wilson J.L., Mikroudis G.K., and Fang H.Y. Expert Systems Application in Environmental Geotechnology. In Beakley G.C., and Haden C.R. (Ed.), *Computer-Aided Processes in Instruction and Research.* ASEE, 1984.

AN EXPERT SYSTEM FOR DIAGNOSIS AND TREATMENT OF DAM SEEPAGE PROBLEMS

M.I. Asgian, [1]AMASCE; K. Arulmoli, [1]*MASCE; W.O. Miller, [2]MASCE; and K. Sanjeevan, [1]AMASCE

ABSTRACT

This paper describes the expert system EXSEL for diagnosing seepage problems associated with dams. The expert system is a 'user-friendly' computer program which runs on a 512K IBM XT compatible personal computer. The user is prompted for a description of a seepage problem, is given a tentative diagnosis, and is presented with some possible remedial actions. The expert system also allows the user to access a data base containing case histories of dams which have had major seepage problems, and thus, provides the user with information about ways in which others solved seepage problems similar to the one at hand. The expert system was written primarily for civil engineers and geologists.

INTRODUCTION

The problem of seepage/leakage through dams has been and continues to be a challenge to the civil engineers and geologists involved in the construction and maintenance of hydraulic structures. The evidence for this is ample. A four year long study of non-federal dams revealed that of the 8800 or so structures inspected, about 28 percent were found to have seepage and leakage as serious concerns (Department of the Army, 1982). Over 15 percent of the dams that are operated and maintained by the U.S. Army Corps of Engineers have been identified as having, or have had, seepage/leakage as a concern (Arulmoli, 1986).

Any successful solution to a seepage/leakage related problem hinges mainly on the experience of the team undertaking the task of remedying that problem. An expert system can provide a means of utilizing the knowledge of dam experts to provide solutions to a given seepage problem. The expert system EXSEL (Expert System for diagnosing SEepage and Leakage associated with dams) was developed in an effort to achieve this goal.

---

[1] J.F.T. Agapito and Associates, Inc., 715 Horizon Drive, Suite 340 Grand Junction, CO 81506   *(currently at The Earth Technology Corporation, 3777 Long Beach Blvd., Long Beach, CA 90807)
[2] Waterways Experiment Station, P.O. Box 631, Vicksburg, MS 39180

CAPABILITIES OF THE EXPERT SYSTEM

EXSEL has been constructed primarily as a diagnostic tool. As such, it is programmed to query the user about the relevant symptoms of dam seepage problems. The user provides the expert system with a case history during a question and answer session. The questions are multiple choice questions which may have multiple answers. Once the expert system has been provided with the symptoms of a problem, it determines the likely cause(s) of the problem and recommends possible remedial measures which can be taken.

EXSEL deals only with qualitative information (e.g., high piezometric levels, presence of wet spots, turbid seepage, change in flow rate, etc.), and thus, is valuable for making preliminary assessments of seepage problems. Final assessments require quantitative information (piezometric levels, flow rates, permeabilities, etc.), and they require the expertise of experienced engineers and geologists.

EXSEL uses the expert system shell ARITY PROLOG (Arity Corporation, 1986) to prompt the user for a description of the seepage problem. The shell uses the backward chaining technique to manipulate the knowledge base and arrive at a diagnosis. The knowledge base consists of both rules (IF - THEN statements) and frames (a set of interrelated objects which provide a shorthand way of programming the rules). The reader is referred to Harmon and King (1985) and to Waterman (1986) for detailed discussions on frames and rules.

RULES CONTAINED IN THE EXPERT SYSTEM

The rules contained in the knowledge base were constructed in two stages. A preliminary set was based upon the expertise of the authors, and upon information given in the literature, especially from the publication of the National Research Council (1986). The preliminary set of rules was presented to a panel of experts for review. All of the preliminary rules pertaining to earth dams (72 out of a total of 80) were reviewed by experts during a 3-day panel discussion which was held at the Waterways Experiment Station in Vicksburg, Mississippi.

The rules are categorized according to primary manifestations which the dam seepage problems can take. The eight primary manifestations of seepage considered in the rules are as follows:

1. Turbid seepage or seepage carrying fines
2. Localized seepage/wet spots/soft or quick spots
3. High piezometric levels
4. Boils
5. Change in flow rate in drains
6. Presence of holes or depressions
7. Whirlpool in reservoir
8. Mass movement (slides, slumps, bulges, cracks, etc.)

Some of these manifestations of seepage problems also can be manifestations of other types of problems. For example, mass movement can be due to excess pore pressure and it could be due to earthquake loading. To discern whether or not a problem is a seepage problem, the expert system questions the user about other symptoms and observations (e.g., occurrence of a recent earthquake, presence of wet spots on embankment, unusually high piezometric readings, etc.). Based on the additional observations, the expert system determines if the problem at hand matches any of the problems described by the rules. If one or more of the rules is applicable, the expert system informs the user that the likely cause(s) of the problem have been determined. Then, the expert system proceeds to list the causes. Some of the causes of seepage problems which the expert system takes into consideration include the following:

   Inadequate filter design
   Clogged drains or filters
   Clogged relief wells
   Development of cracks due to hydraulic fracturing
   Dispersive soils
   Piping because of inadequate design/construction
      (e.g., slope too steep, lenses of improper material,
      absence of zonation, etc.)
   Excessive piezometric gradients in foundation
   Dissolution of soluble materials
   Breach in blanket or cutoff
   Inadequate core/cutoff/foundation preparation

Some remedial measures which can be taken to alleviate these problems are also contained in the set of rules. Potential remedial measures which are suitable for the problem(s) are hand presented to the user.

A listing of all the rules describing symptoms, causes, and remedies of dam seepage problems is contained in the report by Asgian et al. (1987). An example of one rule contained in the expert system is given below. It pertains to turbid seepage emerging from the embankment of an earth dam:

   IF   (the dam is an earth dam, and
         the turbid seepage is coming from the downstream slope of
            the embankment, and
         the turbid seepage is not associated with a structural
            feature such as a spillway wingwall, and
         there is no distress (cracking) in the dam, and
         the piezometric levels are normal)

   THEN (the problem may be due to piping because of inadequate
         design/construction, e.g., too steep slope, lenses of
         improper material, absence of zonation, etc. Potential
         remedial measures which can be taken include the following:
            Monitor
            Partial reduction of reservoir pool level
            Install inverted filter
            Toe drain

Filter berm
Cutoff
Please consult an expert to determine which one is most suitable for your problem.)

A problem matching this rule or any of the other rules contained in the expert system, can be diagnosed during a question and answer session with the expert system.

CONSULTING THE EXPERT SYSTEM

The expert system prompts the user for a description of the problem at hand by asking questions and offering possible answers in menu form: the user selects the most appropriate answer(s) for each question. The user is prompted for all the information needed to make a tentative diagnosis of a dam seepage problem. Once all the pertinent information is collected, the expert system diagnoses the problem and provides a list of potential remedies. A sample session with the expert system is given in Figure 1. The session diagnoses the turbid seepage problem described in the previous section.

Upon diagnosis of the problem, a report is automatically generated and the user is given the option to reconsult the expert system. Although not shown in the sample session, the user is also given the option to access a data base containing case histories of dam seepage problems.

CONSULTING THE SEEPAGE DATA BASE

A data base on dam seepage problems was constructed by Arulmoli (1986) during a previous, closely related study. The data base contains case histories of 115 dams which are maintained by the U.S. Army Corps of Engineers: each of the dams has had one or more seepage problems. By consulting the data base, the user can read case histories of dams which have had problems similar to the one at hand. The case histories contain detailed information regarding the manifestation of the seepage problem, the geological setting, the dam construction, the original remedial seepage control measure(s) taken, and the efficacy of the original seepage control measure(s).

As with the expert system, information from the data base is retrieved during a question and answer session. The user is presented with a menu of different categories of case histories. The user can selectively choose to examine case histories by problem type (e.g., wet spots, turbid seepage, abnormally high piezometric readings, etc.) by dam name, by U.S. Army Corps of Engineer district number, or by the state in which the dams reside. A report can be generated which contains the case histories of interest.

```
*                  Welcome to EXSEL                    *
*                                                      *
*                An Expert System to                   *
*      Diagnose and Recommend Remedial Actions for     *
*           Seepage-Leakage Problems in Dams           *
*                                                      *
*                   Developed by                       *
*              J.F.T. Agapito & Associates             *
*              27520 Hawthorne Blvd, Suite 295         *
*              Rolling Hills Estates, CA  90274        *
*              (213)544-0474 and (303)242-4220         *
*                        and                           *
*     U.S. Army Engineers, Waterways Experiment Station*
*                    Vicksburg, MS  39180              *
*                                                      *
*                    Version 1.0                       *
*                    August 1987                       *
********************************************************

                          continue
                          explain


1. What type of dam is being considered?

       1) Earth embankment dam
       2) Concrete or roller compacted concrete dam
       3) Rockfill dam
       4) why ?
       5) unknown



2. Do you see any of the following primary
   manifestations of seepage?(can be anywhere- not just on
   the embankment)

       1) Boils
       2) Evidence of the following damage to appurtenent structures:
          a)leaky or stained conduit
          b)conduit carrying soil deposits
          c)cracked or misaligned structures such as the spillway or
            stilling basin
       3) Changes in flow rates in toe drainage system
       4) Heave of foundation near the toe
       5) Abnormally high piezometric readings
       6) Holes/Depressions
       7) Signs of mass movement of the embankment or abutment
          (e.g., slides,sloughs,bulges,sag in crest,cracks,misaligned
          guard rails etc.)?
       8) Turbid seepage or seepage carrying fines
       9) Wet/Soft/Quick spots or marshy areas
      10) Whirlpool in reservoir
      11) why ?
      12) unknown

? 8
```

Figure 1. Sample Session with EXSEL

3. Is the reservoir empty?

   1) yes
   2) no
   3) why ?
   4) unknown

? 2

4. Is there turbid seepage on the embankment
   (e.g. on the downstream slope, in the toe drainage system
   and at the abutment/embankment interface)?

   1) yes
   2) no
   3) why ?
   4) unknown

? 1

5. Is turbid seepage emerging from the
   downstream slope of the embankment?

   1) yes
   2) no
   3) why ?
   4) unknown

? 1

6. Is the turbid seepage on the embankment
   randomly located?(i.e.,NOT associated with the abutment/embankment
   interface or any other structural feature)

   1) yes
   2) no
   3) why ?
   4) unknown

? 1

7. Are the piezometric levels in the embankment
   unusually high?

   1) yes
   2) no
   3) why ?
   4) unknown

? 2

8. Do you see one or more of the following?

   1) Transverse shrinkage or settlement crack(s) on the embankment
   2) Longitudinal cracks on the embankment
   3) Arc-shaped crack(s) on the embankment
   4) None of the above
   5) why ?
   6) unknown

? 4

9. Is turbid seepage emerging through the
   drainage system?

   1) yes
   2) no
   3) why ?
   4) unknown

? 2

Figure 1. Sample Session with EXSEL (Cont'd)

```
10. Is turbid seepage emerging from a hole
    or burrow?

            1) yes
            2) no
            3) why ?
            4) unknown
? 2


11. Is turbid seepage emerging from or
    around the perimeter of the conduit?

            1) yes
            2) no
            3) why ?
            4) unknown
? 2


12. Is turbid seepage adjacent to the
    spillway or stilling basin?

            1) yes
            2) no
            3) why ?
            4) unknown
? 2


13. Is there turbid seepage on the foundation
    (i.e., on the toe, in relief wells, and in the general vicinity
    of the dam)?

            1) yes
            2) no
            3) why ?
            4) unknown
? 2


****************************************************************

The turbid seepage on the downstream slope may be
due to piping because of inadequate design/construction(e.g., too steep slope,
lenses of improper material, absence of zonation, etc.)

Possible remedies include one or more of the following items.
Please consult an expert for most suitable remedies for your particular dam.
        Monitor.
        Partial reduction in reservoir pool level.
        Toe drain.
        Filtered pipe collector system.
        Filter berm.
        Cutoff.

Conditions which lead to this conclusion:
        the dam is an earth dam,
        and there is turbid seepage on the downstream slope,
        and the seepage is NOT associated with a structural feature,
        (e.g., abutment/embankment interface, conduit, etc.)
        and there is NO distress (cracking) in the dam,
        and the piezometric levels are normal.

-- More --

Would you like another consultation ?n.
```

Figure 1. Sample Session with EXSEL (Cont'd)

SOFTWARE AND HARDWARE REQUIREMENTS

The expert system EXSEL is driven by the shell ARITY PROLOG. An unlimited number of copies of the executable code can be distributed without any licensing fee. However, any user wishing to make changes to the code must purchase the necessary software (the expert system shell, and the interpreter and compiler) from the Arity Corporation.

As a stand-alone program, the expert system runs on a 512K IBM XT compatible personal computer. If the seepage data base is consulted in conjunction with the expert system, then a 640K IBM compatible personal computer is required.

The seepage data base is accessible through the data management computer program dBASE III marketed by Ashton-Tate. Purchase of the dBASE III software is required to access the information contained in the data base.

CONCLUSIONS

An expert system has been constructed as a diagnostic tool for seepage problems associated with earth dams, rockfill dams, concrete dams and roller compacted concrete dams. The expert system is a 'user-friendly' computer program which prompts the user for a description of the seepage problem. Based on the information provided by the user, the expert system determines the likely cause(s) of the problem and suggests potential remedial actions. The expert system also allows the user to access case histories of dams which have had seepage problems similar to the one at hand.

ACKNOWLEDGEMENTS

Development of the expert system EXSEL was funded by the U.S. Army Engineers Waterways Experiment Station under the REMR-2 Program (Repair, Evaluation, Maintenance, and Rehabilitation Research Program). The panel of experts, Drs. James Erwin and Thomas F. Wolff, and Messrs. Robert L. James, Ben Kelly, and Lloyd B. Underwood, provided much constructive criticism of the set of rules. Dr. Christopher St. John provided guidance in the development of the expert system.

APPENDIX I.- REFERENCES

Arity Corporation (1986) 'The Arity/Expert Development Package,' Arity Corporation, Concord, MA.

Arulmoli, K. (1986) 'Compilation of Database on Remedial Measures to Control Seepage and Leakage through Corps of Engineers Civil Work Structures,' Report prepared by J.F.T. Agapito and Associates, Inc. for the U.S. Army Engineers Waterways Experiment Station, Vicksburg, MS under Contract No. DACA39-85-C-0021, August.

Asgian, M.I., Arulmoli, K. and K. Sanjeevan (1987) 'Expert System for Diagnosing Seepage Problems Associated with Dams,' Report prepared by

J.F.T. Agapito and Associates, Inc. for the U.S. Army Engineers Waterways Experiment Station under Contract No. DACW39-86-C-0052, August.

Department of the Army, Office of the Chief of Engineers (1982) 'National Program of Inspection of Non-Federal Dams,' Final report to Congress. Washington, D.C., May.

Harmon, Paul and David King (1985) Expert Systems, Artificial Intelligence in Business, John Wiley & Sons, Inc., New York.

National Research Council (U.S.) Committee on the Safety of Existing Dams (1986) 'Safety of Existing Dams,' Third Printing. National Academy of Sciences, Washington, D.C.

Waterman, D.A. (1986) A guide to Expert Systems, Addison-Wesley Publishing Company, Inc., Reading, Massachusetts.

ASA: AN EXPERT SYSTEM FOR ACTIVATED SLUDGE ANALYSIS

D. G. Parker[*], M. ASCE, and S. C. Parker[**]

Abstract

ASA, the expert system described in this paper, is designed to provide assistance to a consulting engineer or to the personnel charged with the operation and control of an activated sludge treatment process. The activated sludge process is capable of producing a high quality effluent, however, proper operation and control is essential to achieve optimum performance. Studies have shown that most problems with activated sludge systems are due to faulty operation rather than faulty design.

ASA is designed to diagnose and/or analyze a problem or problems that develop with the operation and control of an activated sludge treatment process. Once a problem or problems are identified, the system directs the user in correcting the situation and bringing the process under control. The operator must be familiar with the operation procedures and laboratory procedures for the process but does not need to be expert in diagnosing problems in the process in order to use the system.

Programming for ASA was developed using the Borland International Turbo programming language. This language has many features which make it ideally suited for programming an expert system.

Introduction

The class of computer programs called expert systems represent a branch of artificial intelligence. Artificial intelligence is the branch of computer science that "is primarily concerned with knowledge representation, problem solving, learning robotics, and the development of computers that can speak and understand more natural (humanlike) languages" (Townsend 1986). Expert systems are

---

[*]Professor, Department of Civil Engineering, 4190 Bell Engineering Center, University of Arkansas, Fayetteville, AR 72701.
[**]Associate Professor, Department of Industrial Engineering, 4207 Bell Engineering Center, University of Arkansas, Fayetteville, AR 72701.

artificial intelligence hardware and/or software systems which use deductive reasoning to solve problems in specific disciplines. Such systems solve problems that usually require significant human expertise for solution, and these systems use knowledge and inference procedures in their solutions.

In designing and implementing an expert system, a knowledge engineer works with appropriate experts and the pertinent literature in the particular field to capture the relevant data and relationships that the experts use to solve problems. The architecture of an expert system, thus, consists of a database that is structured in such a way that it can be used to solve a problem by deductive reasoning, emulating the process by which the human expert would solve the problem.

In the design of ASA, the knowledge base was provided by D. G. Parker, a Civil Engineer who has worked with treatment plant personnel on sludge settling problems. S. C. Parker, an Industrial Engineer with experience in developing expert systems, served as the knowledge engineer.

ASA: An Expert System for Activated Sludge Operation and Control

ASA, the expert system described in this paper, is designed to provide assistance to a consulting engineer or to the personnel charged with the operation and control of an activated sludge treatment process. The most commonly used secondary wastewater treatment process is the activated sludge process. One reason for its wide application is that the activated sludge process is capable of producing a high quality effluent; however, proper operation and control is essential to achieve optimum performance. Studies have shown that most problems with activated sludge systems are due to faulty operation rather than faulty design (EPA-625/6-84-008 1982).

The most common problem with activated sludge operation is poor settling in the final clarifier. Activated sludge settles well in a properly designed clarifier when the environmental conditions in the plant are such that the microorganisms in the mixed liquor form large, relatively dense flocs. Settling problems in activated sludge, however, can be very complex. Many design engineers and treatment plant operators do not have sufficient knowledge of the interaction of environmental conditions, microbiology, and floc formation to solve some of the more complex sludge settling problems. In fact, experts in the field may not be able to quickly diagnose some settling problems. Although there is a considerable body of expert knowledge concerning activated sludge settling problems, there are some incidences in which diagnosis is difficult and depends on expert judgement and experience. In some cases, a problem cannot be uniquely identified by laboratory and field tests, and, in fact, more than one problem can be occurring simultaneously. Furthermore, not all treatment plants respond the same to a given set of circumstances. In complex cases, it may be necessary to implement corrective procedures in sequence or parallel until the problem or problems are solved and the process brought under control. Given certain characteristics to a problem, ASA is designed to use expert judgement in proceeding with sequential corrective measures when the settling problem is complex.

The System Program

Programming for the system was developed using the Borland International Turbo Prolog language (Borland 1986). A prototype of parts of the system was developed using EXSYS, an expert system shell (EXSYS 1985). EXSYS is a rule-based system and was convenient and relatively easy to use for testing some of the initial ideas for building the system. However, it was decided to use a logical programming language for the actual system development because of the greater flexibility afforded by a logical language.

Turbo Prolog was chosen from the possible logical languages for a variety of reasons. Primary among these reasons is its wide availability. The cost is low ($99), and it runs on an IBM PC or compatible machine, making it readily accessible to even the smallest wastewater treatment plant.

There are several programming advantages to Turbo Prolog. It has a powerful and flexible built-in editor with commands similar to those of WordStar (Shafer 1986). This feature accommodates the programming stage of system development. In addition, the language is supplied with a menu routine, named menu.pro (Townsend 1987). Menu.pro may be used to produce a menu function automatically, when queries are such that a question has one of several alternative answers. Another useful function of Turbo Prolog is the ability to build and save a dynamic database (Borland 1986) during program execution. As information is collected from the user, it may be stored in a dynamic database. This database may then be accessed during the same diagnostic session, thereby preventing the system from querying the user for data already presented in earlier queries. The dynamic database, thus, serves to 'remember' data acquired by the system until the end of a diagnostic session, after which the database is cleared for use in the next session.

Turbo Prolog is not a procedural language, but is instead an object-oriented language (Townsend 1987). A Prolog program, consisting of a collection of data or facts and the relationships among these facts, is able to infer facts and conclusions from other facts. The collection of facts and relationships making up such a program constitute rules which allow inferences to be made. The execution of such a program does not follow a particular procedure, but is controlled by symbolic processing which is influenced by the particular responses of the user in a diagnostic session. A Turbo Prolog program is designed to construct an hypothesis (the wastewater problem diagnosis for example) and, using formal reasoning and backtracking, attempt to prove or disprove the hypothesis based on the known and provided data.

The System Design

The general design of ASA, the activated sludge wastewater treatment expert system, is illustrated in Figure 1. The user, in this case the consulting engineer

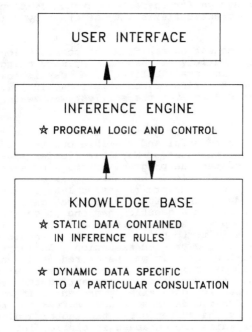

Figure 1. General form of ASA, expert system for Activated Sludge Analysis

or the treatment plant operator, interfaces with the
program in a diagnostic consultation by providing
descriptive information and results of laboratory analyses.
The program is equipped with an 'inference engine', which
is made up of the statements in the program which provide
the logic and control of program operation. The knowledge
base with which the program logic works is composed of two
components: a static database consisting of expert
knowledge which is contained in the inference rules built
into the program, and a dynamic database which consists of
the information provided by the user during a diagnostic
session.

In the design of this expert system, the knowledge
engineer and the wastewater treatment expert have taken
into consideration the possible problems that may arise in
an activated sludge treatment process. The knowledge in
the knowledge base and the inferences built into the
system concerning this knowledge have been obtained from
experience as well as from information found in the
literature regarding the activated sludge process (EPA
430/9-77-006; Benefield and Randall 1980; Gaudy and Gaudy
1988).

The operational problems addressed by this system are
filamentous bulking sludge, floating sludge, ashing,
various flocculation problems, and possible combinations of
these problems. The system is designed to handle problem
solutions which are essentially algorithmic in nature, for
example, the diagnosis and treatment of filamentous bulking
sludge when this problem occurs alone, as well as those
which require judgement and heuristic solution approaches.
In the latter type of problem, several causative elements
may be present simultaneously, rendering a positive
diagnosis impossible. The approach that is taken by the
system is a heuristic solution technique which consists of
making sequential corrections to the wastewater system in
an order based on expert judgement, assessing the effects
of the corrections, and making additional corrections as
deemed necessary.

The general form and flow of a diagnostic session with
this expert system is illustrated by Figure 2. The
program user is first requested to provide information
about the problem in terms of observations regarding the
clarifier and/or settling test routinely run by treatment
plant laboratories. The system diagnoses the class of
problem (for example, bulking sludge, ashing, flocculation
problems, and so on). Based on diagnosis at this level,
the system requests more specific information about the
results of specific lab tests or process parameters. This
latter information allows the system to either definitely
identify the problem or to identify two or three different
conditions which could be causing the problem. In the
latter case, it is possible for more than one of these

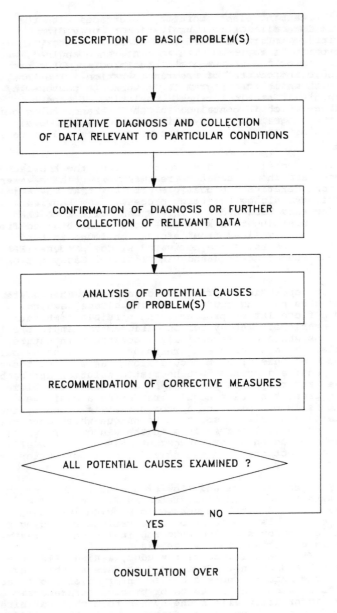

Figure 2. Form and flow of a consultation with ASA, expert system for Activated Sludge Analysis.

conditions to be present simultaneously. At this point, the system requests additional information concerning operating parameters, and, on the basis of the responses, solutions are suggested to correct the problem or problems. In the case of multiple possible problems without a definite diagnosis, the system is designed to suggest corrective measures in a particular order as defined by the expert. This process is heuristic in nature, since the results of the attempted solution are analyzed to determine whether other corrections are necessary. Through an iterative procedure, the process is brought under control.

Example Diagnostic Session with Definite Diagnosis

An activated sludge wastewater treatment problem that is straight forward to diagnose is filamentous bulking sludge. The following illustrates the interaction between the user and the expert system when this is, in fact, the only problem at hand.

The initiation of the program fires the presentation of a screen which requests information from the user concerning the general description of the problem conditions. This request is in the form of a menu with four different descriptions of clarifier conditions and instructions to the user to select the statement which best fits the problem situation. If, as in our example, the problem is filamentous bulking sludge, the user will choose the following statement as being descriptive of the problem:

"In the clarifier the sludge blanket is too high with normal recirculation and/or in the settling test the SVI is over 150 ml/g."

The user will then be informed on a second screen that the problem is tentatively diagnosed as bulking sludge. In effect the program has invoked a rule which could be stated as follows:

IF   IN THE CLARIFIER THE SLUDGE BLANKET IS TOO HIGH
           WITH NORMAL RECIRCULATION,
AND/OR   IN THE SETTLING TEST THE SVI IS OVER 150 ml/g,
THEN   THE PROBABLE PROBLEM IS BULKING SLUDGE.

The second screen will state also that in order to confirm this diagnosis, a microscopic examination of the sludge will be necessary. The user is requested to respond to the following question:

"The sludge flocs contain an excessive quantity of filaments (y/n) ?"

Given the problem of filamentous bulking sludge, the user will respond positively to this question. This positive response will fire a third screen which in this case will consist of a positive diagnosis of filamentous bulking sludge, with the note that excessive filaments prevent the settled sludge from compacting sufficiently. On this screen also, the user will be requested, in the form of a menu, to select the statement about process parameters and wastewater characteristics which is correct in this situation. The user may select the following statement as describing the parameters of the operation:

"The influent to the aeration tank has a nitrogen to BOD_5 ratio less than 0.05."

In the event that this is the statement chosen, the indication is that a deficiency of nitrogen is the cause of the problem. A fourth screen is presented with information concerning the cause of the problem and a request for additional parameter values. After the user inputs the appropriate parameter values, a fifth screen is fired which presents the calculated values for the deficiency. In effect an expanded rule has been invoked which could be stated as follows:

IF	IN THE CLARIFIER THE SLUDGE BLANKET IS TOO HIGH WITH NORMAL RECIRCULATION,
AND/OR	IN THE SETTLING TEST THE SVI IS OVER 150 ml/g,
AND	THE SLUDGE FLOCS CONTAIN AN EXCESSIVE QUANTITY OF FILAMENTS,
AND	THE INFLUENT TO THE AERATION TANK HAS A NITROGEN TO BOD_5 RATIO LESS THAN 0.05,
THEN	AT LEAST ONE CAUSE OF THE BULKING SLUDGE PROBLEM IS DEFICIENCY OF NITROGEN IN THE PROCESS.

The fourth screen will appear as follows, illustrating the interaction with the user to collect parameter values and present calculated values for the deficiency:

Filamentous organisms tend to grow more rapidly than floc-forming bacteria in activated sludge when an insufficient supply of nitrogen is available. The actual quantity of nitrogen required is a function of the quantity of sludge wasted.

Enter the requested number and press return:

Sludge wasting rate (pounds of VSS per day)? 1500*

Influent nitrogen (pounds per day)? 160*

---
*Values input by user

The fifth screen appears with calculated values for the nitrogen deficiency and information concerning corrective measures:

Calculated deficiencies are as follows:

    Nitrogen deficiency (pounds per day) = 23*

    Nitrogen deficiency (percent) = 20*

Nitrogen requirement can be reduced by reducing the sludge wasting rate. If sludge wasting can be reduced by as much as the percent of nitrogen deficiency, then no nitrogen addition will be required.

Nitrogen can be added in the form of anhydrous ammonia or other nitrogen sources to the influent to the aeration tank. Supplying sufficient nitrogen to the aeration tank should result in a gradual reduction in the quantity of filaments.

For a quick reduction in the filiment concentration, chlorinate the return sludge at a rate of 2-3 pounds of chlorine per day per 1000 pounds of MLVSS. Chlorination should be considered only a short term solution and should not be continued if the above methods produce satisfactory results.

---

*Values calculated by the program

It is possible that other conditions exist which are also influencing the sludge to bulk. For this reason, the program is designed to allow the user to return to the menu of the third screen which presents statements that describe the process parameters. The user may choose another statement as being also correct concerning the present conditions of treatment plant operation, and the program will proceed to further analyze this problem situation and provide corrective measures. Any problems which exist may be so analyzed by the user of the system.

Flocculation Problems

    The above procedure is descriptive of the system operation when the treatment plant problems are diagnosed as bulking sludge, floating sludge, or ashing. There are, however, frequently problems of flocculation which can be quite complicated and difficult to diagnose. There are three conditions, food to microorganism ratio out of range, dissolved oxygen too low, and toxicity level too high, that may be the cause of flocculation problems. The conditions regarding protozoa in the sludge is a useful indicator in diagnosing these flocculation problems. Tables 1 and 2 illustrate the various levels of these conditions, and the problem cause or causes of the flocculation problem given these levels. For some combinations of these parameters, a single cause can be identified, however, other combinations may be indicative of one to three possible causative factors. The particular values that constitute high and low levels of food to microorganism ratio are a function of the characteristics of a particular treatment plant. The following abbreviations are used in the tables:

P1 - Protozoa are present and active
P2 - Protozoa are present but not active
P3 - Protozoa are absent

D1 - Dissolved oxygen level is greater than or equal to 2
D2 - Dissolved oxygen level is less 2

F1 - Food to microorganism ratio is high
F2 - Food to microorganism ratio is normal
F3 - Food to microorganism ratio is low

ASA, the expert system, is designed to collect the information about the above three wastewater characteristics in three separate menus. The probable cause or causes of the flocculation problem are then reported to the user, along with corrective measures.

Table 1. Causative factors for flocculation problems when dissolved oxygen is greater than or equal to 2.

	D1		
	P1	P2	P3
F1	F1	F1 and/or toxicity	F1 and/or toxicity
F2	should have no operational problem	toxicity	toxicity
F3	F3	F3 and/or toxicity	F3 and/or toxicity

Table 2. Causative factors for flocculation problems when dissolved oxygen is less than 2.

	D2		
	P1	P2	P3
F1	F1 and/or D2	F1 and/or D2 and/or toxicity	F1 and/or D2 and/or toxicity
F2	D2	D2 and/or toxicity	D2 and/or toxicity
F3	F3 and/or D2	F3 and/or D2 and/or toxicity	F3 and/or D2 and/or toxicity

In the event of a single probable cause, the diagnose is highly likely, and the necessary corrective measure is clearly indicated. However, in the event of the possibility of two or more causative factors, or a situation where different causative factors cannot be distinguished at this level of information, the system will lead the user through a series of corrective measures, analyzing the results after each one, and deciding if further action is needed.

Conclusion

The expert system described in this paper is designed to be useful to the activated sludge treatment plant operator and the consulting engineer, and it may serve as an educational tool as well. It makes use of a database built into the inference statements of the program and collects data from the user during a consultation that are pertinent to the problem at hand. These data are analyzed, problems diagnosed, and corrective measures suggested to allow the settling problems to be controlled.

Appendix. - References

Benefield, L. D., and Randall, C. W. (1980). *Biological Process Design for Wastewater Treatment*, Prentice-Hall, Inc., Englewood Cliffs, NJ.
Borland (1986). *Turbo Prolog: The Natural Language of Artificial Intelligence*, Borland International, Inc., Scotts Valley, CA.
EXSYS (1985). *EXSYS, Version 3, Expert System Development Software*, EXSYS, Inc., Albuquerque.

Gaudy, A.F., and Gaudy, E. T. (1988). <u>Elements of Bioenvironmental Engineering</u>, Engineering Press, Inc., San Jose.

<u>Handbook for Improving POTW Performance Using the Composite Correction Program Approach</u> (1982). EPA-625/6-84-008, U. S. EnvironmentalProtection Agency, Center for Environmental Research Information, Cincinnati.

<u>Process Control Manual for Aerobic Biological Wastewater Treatment Facilities</u> (1977). EPA 430/9-77-006, U. S. Environmental Protection Agency, Municipal Operations Branch, Office of Water Program Operations, Washington, D.C.

Robinson, P. R. (1987). <u>Using Turbo Prolog</u>, Osborne McGraw-Hill, Berkeley.

Shafer, Dan (1986). <u>Turbo Prolog Primer</u>, Howard W. Sams & Co., Indianapolis.

Townsend, C. (1987). <u>Introduction to Turbo Prolog</u>, Sybex, Berkeley.

PART III

CONSTRUCTION

KNOWLEDGE ELICITATION TECHNIQUES
FOR CONSTRUCTION SCHEDULING[1]

by

Jesus M. De La Garza[2], C. William Ibbs,[3] M. ASCE, E. William East,[4] M. ASCE

ABSTRACT

Knowledge elicitation is perhaps the most ambitious, important, time-consuming, ill-structured, and challenging phase of a knowledge engineering project. Its purpose here is to determine the practical feasibility of capturing construction scheduling knowledge for a Knowledge-Based System (KBS). This objective was accomplished by investigating knowledge acquisition methodologies capable of stimulating and structuring what initially appeared to be an amorphous mass of knowledge, while not getting diverted by the specifics of implementation.

The lessons learned during this process as well as the structured expertise represent major contributions to the construction industry.

INTRODUCTION

Construction scheduling diagnosis is controlled through expert knowledge, judgment, experience, and traditional CPM/PERT computational techniques. The process often lacks structure, and is tedious, repetitive, and time consuming; thus, a thorough analysis of all factors potentially affecting and explaining changes in the schedule is not always possible. Nevertheless, field engineers do recognize that they could perform more meaningful analyses as well as react more quickly to changed conditions with tools that free them from the burden of routine and cumbersome tasks and tools that can sustain a high level question/answer dialogue.

The U.S. Army Construction Engineering Research Laboratory (USA-CERL) and the University of Illinois Construction Engineering Expert Systems Laboratory (CEESL) have been working together to develop a knowledge-based

---

[1] Presented at the Microcomputer Knowledge-Based Expert Systems in Civil Engineering Symposium, during the ASCE National Convention in Nashville, Tennessee, May 1988.

[2] Ph.D. Research Assistant, Department of Civil Engineering, University of Illinois at Urbana-Champaign, Illinois, 61801.

[3] Associate Professor, Department of Civil Engineering, University of California at Berkeley, California, 94720.

[4] Principal Investigator, U.S. Army Construction Engineering Research Laboratory, Champaign, Illinois, 61820.

system for analysis of construction schedules [De La Garza 88]. Schedule analysis and evaluation are divided into two areas, namely an Initial schedule analysis module and an In-Progress schedule analysis module. Each is based upon four major subcategories: (a) cost; (b) time; (c) logic; and (d) general requirements. This paper describes the knowledge acquisition techniques employed in this knowledge engineering effort.

ISSUES AND METHODS

The knowledge engineer works hand in hand with a domain specialist whose knowledge is desired to transfer. The knowledge engineer attempts to drain the required expertise from the experts' head in the form of rules which can be encoded in a machine system. There is no single methodology for the process of knowledge elicitation that has proved universally effective. In this section, a working classification of methods for eliciting experts' knowledge is outlined [Hart 85, Hoffman 87, Trimble 86].

Unstructured interview. The knowledge engineer does not have a list of detailed questions to ask. Rather, he or she asks spontaneous conceptual questions while the expert talks about the domain. For example, the knowledge engineer might ask the expert a question such as "What constitutes good scheduling practices?" This technique can involve making an audiotape of the expert's discourse and producing a transcript. The technique produces a lot of relevant and irrelevant information, whose analysis can be fairly time-consuming.

Method of familiar tasks. This technique involves observing the expert while he or she is at work. Studying the experts implies the identification of goals, sub-goals, data to which they like to have access, and the information they produce. This method is usually applied to generate the first pass at the knowledge base. It also includes the collection and analysis of relevant public domain knowledge.

Structured interview. In this technique, the knowledge engineer utilizes the knowledge base developed by using the "method of familiar tasks" or the "unstructured interview". The expert comments on each entry in this knowledge base to confirm it, expand it, delete it, qualify it, or reorganize it. The knowledge engineer switches from general to more specific questions like "How do you know that the logic of this schedule would not work?" The product of this method can be considered a second generation knowledge base.

Limited information tasks. This technique tinkers with the thought process by restricting the amount and kind of information available to the expert otherwise. This technique is useful to stimulate the formulation of hypothesis and strategic thinking. Because the method forces the expert to heavily rely on compiled and intuitive knowledge, this technique is likely to frustrate the expert and make him or her feel uncomfortable.

Constrained processing tasks. This technique differs from the "limited information tasks" in that it imposes external constraints on the expert. That is, it either a) limits the time within which the expert learns factual data and draws recommendations; or b) asks the expert a specific question rather than a full analysis of a situation. As with the "limited information tasks" technique, experts might be hesitant to give advice in an unfriendly environment.

Method of tough cases. This technique attempts to unveil subtle aspects of the expert's thought process. These aspects may be stimulated during the analysis of a case that is either too simple or too complex.

Induction. Induction, the opposite of deduction, encourages an expert to refer to specific examples of cases which consist of decisions and parameters considered in reaching those decisions. A computer program iterates on this set of positive examples to induce general rules that cover at least the training set of examples. The efficiency of the induced rules lies in the selection of the relevant parameters and the training set of examples.

SOURCES OF EXPERTISE

Since an objective of this research is to generate a knowledge base for the development of a knowledge-based system, as opposed to an expert system, the human experts do not need to be the very best persons in the field, but rather, individuals with enough knowledge to solve problems well. In fact, this is actually the case for so many knowledge-based systems in use today.

The nature of the Construction Industry is such that the knowledge about construction scheduling, even for the same type of project, is unlikely to be found in a single human expert. Thus, it was important to find and fuse knowledge from multiple experts. The advantages of using a diverse collection of experts, as outlined by [Mittal 85], more than offset the extra time required to resolve conflicts and contradictions generated by multiple inputs. In fact, the contradictions may be turned into advantages as they indicate areas for further research.

The sources of construction scheduling expertise can be categorized in four groups: a) Contractors; b) Owners; c) Consultants; and d) In-house. W.E. O'Neil Construction Company, a large building contractor in Chicago Illinois, collaborated in this knowledge engineering project by designating one senior project manager who committed sufficient time to the development of the system. Representatives from the USA-CERL and the US Army Corps of Engineers participated to articulate an owner's view. A consultant from Pinnell Engineering, Inc., who provides scheduling services to both owners and contractors, also took part in the knowledge elicitation process. Finally, the in-house expertise of several faculty members in the Civil Engineering Dept. was drawn upon to contribute to the refinement and extension of both the contractor's and owner's views.

KNOWLEDGE STRUCTURE

The knowledge base consists primarily of scheduling decision rules, general construction knowledge, and project specific knowledge.

Scheduling decision rules. The scheduling decision rules assure that the construction schedule complies with imposed general requirements, logic, time, and cost constraints. An example of a cost constraint is: cash flow front-end loading is unlawful.

General construction knowledge. The general construction knowledge represents generic expertise that may be applied to a variety of projects. For example, a winter-sensitive activity should not be scheduled in a period when ambient temperatures are expected to be below specified minimums.

Project specific knowledge. This knowledge is idiosyncratic to a particular type of schedule activities. For example, the enforcement of weather-sensitivity constraints on the hammock activity "install drywall" makes it possible to split it into three sub-tasks: metal studs, drywall, and taping.

EMPLOYED TECHNIQUES FOR KNOWLEDGE ACQUISITION

The approach used here for knowledge acquisition was based on successful methodologies adapted from [Freiling 85, Prerau 87] and outlined below.

The techniques for creating the knowledge base fell into three categories. The first category involved analyzing public domain knowledge. This first exploratory step was used to determine the breath and depth of the Construction Schedule Analysis domain. This phase identified whether the Initial and In-Progress schedule analyses, as defined in this research, were sufficiently narrow and self-contained: the aim is not for a knowledge base that is intricately tied to other kinds of knowledge, i.e., automated schedule generation, work package risk identification, cost control, etc. Rather, the goal is to develop a knowledge base that is robust in a limited area within the domain.

The second category of techniques was the interview of experts with regard to knowledge of facts and procedures. The knowledge base produced in the previous phase was used to interact with the experts. By showing this knowledge base to the experts early in the knowledge elicitation process, it was possible to obtain a better understanding of the different kinds of expertise prevalent in the domain and which experts practiced which kind. In addition, the experts involved in the process better understood the scope and complexity of the project.

The third technique involved studying the experts' performance on specific problems. By limiting the information that is typically available to the experts and by constraining the problem experts are to work on, it was possible to de-compiled intuitive knowledge. Following is a description of each of these techniques.

Textbook scheduling knowledge. In order to add structure to subsequent knowledge acquisition efforts, the first pass at a knowledge base was made by analyzing available texts and technical manuals on the subject [Avots 85, Clough 81, Gray 86, O'Connor 82]. Figure 1 shows the evolution of public domain know-how into human say-how. [De La Garza 88] describes this first generation knowledge base, which consists of a categorized list of statements cast in plain English.

Interviews with scheduling experts. The dialogue acquisition technique was utilized to elicit and formalize the experts' knowledge. This method is depicted in Figure 2 as a knowledge engineer's route map. Its purpose is to convert human know-how into human say-how through a process of articulation, of which the expert is supposed to be capable. Once in this form, programming and compiling techniques can convert it into machine code, i.e., machine know-how. At run-time the computer generates the behavior the user expects, i.e., machine show-how.

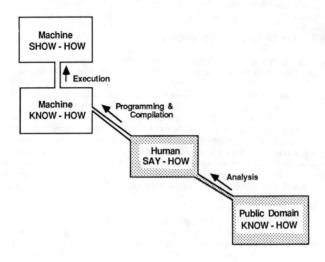

Figure 1. Analysis route map

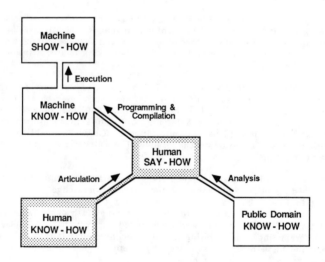

Figure 2. Dialogue route map

In Figure 2 the articulation channel depicts what Artificial Intelligence researchers call the knowledge acquisition problem or the "Feigenbaum bottleneck" [Feigenbaum 83]. The narrowness of the bottleneck is related to the complexity of the task domain. [Michie 86] argues that the more complex the mental skill, the greater the proportion of it which is encoded in intuitive form and hence tends to be beyond the access of the expert.

For the purposes of this research, two experts from W.E. O'Neil Construction Co. were separately interviewed during twelve and four hours respectively. The sixteen hours' worth of interview were divided into four sessions of four hours each.

This technique was proposed and utilized mainly because it was assumed that senior project managers were confident about their ability to access the large body of pattern-based rules which they have in their head. The approach can be categorized as a "Structured Interview" if it is assumed that the knowledge base developed from analysis of technical reports represents a first generation knowledge base. W.E. O'Neil went over this first generation knowledge base one entry at a time making comments on each one. As pointed out by [Hoffman 87], this process led to the addition and deletion of entries, quantification of entries, and addition of categories. [De La Garza 88] explains the knowledge gathered during the interviews.

Although the combination of this approach and already published knowledge produced a more robust series of high level concepts, i.e., the second pass to human say-how, experts often could not describe just how much they know, how they keep track of it, or how they know when to use which information.

Schedule evaluation exercise. A third alternative to elicit knowledge assumes that the route map in Figure 2 is incomplete and that there is some way of going from human know-how to machine know-how other than by the articulation passage. Figure 3 shows a more robust map than its counterpart in Figure 2. In this case, it is possible to proceed from human know-how into tutorial human show-how, that is, human-supplied advice by stimulating the knowledge and reasoning skills of the experts. In addition, it is possible to go from the second generation human say-how to human show-how by manually simulating the application of this knowledge base. After having arrived to this point, a model of the experts' skill in explicit form can then be translated into third generation human say-how rules. Those rules can be subsequently compiled into machine know-how. This approach to knowledge acquisition thus generates a passage to circumvent the articulation bottleneck of Figure 2.

The critical processes for implementing this approach are represented by the bridges that link the human know-how and say-how with the human show-how, i.e., the execution and hand simulation channels. In order to stimulate these channels, this research study included an experiment using two video cameras, a trio of senior project managers, a rookie project manager, and a curtain. The aim was to mimic a computer by having the trio act as the knowledge-based system, the curtain act as the computer screen and the rookie act as the user.

This exercise had the following primary goals: 1) to push the knowledge base toward completeness, by eliciting new knowledge; 2) to validate, fine tune or disprove existing knowledge; 3) to develop a model of interaction between

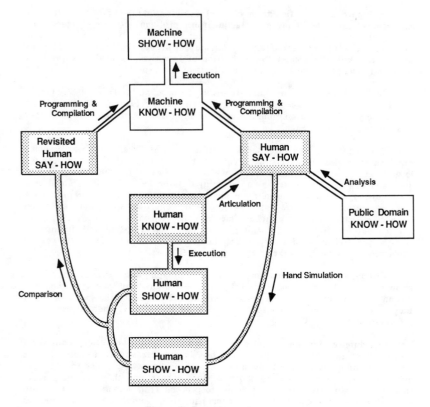

Figure 3. Execute route map

the user and the knowledge-based computer system; and 4) to determine the practical feasibility of using this knowledge elicitation technique to capture construction scheduling knowledge.

The 2-day exercise was divided into five phases: 1) ice breaking; 2) initial schedule analysis; 3) in-progress schedule analysis; 4) overall schedule analysis; and 5) de-briefing.

The ice breaking phase. One of the criteria that clearly was going to affect the outcome of the exercise was the level of interaction among the trio of experts. Thus, the intent in this phase was to introduce the heterogeneous trio members to each other and allow for free interaction. During this period, they were presented with an overview of knowledge-based systems, the objectives of this research project, a detailed description of the mechanics of the exercise, and with a general statement regarding their expected contribution. They also had the opportunity to get acquainted with the

Banyan^{Tm} Local Area Network. A total of four hours were devoted to this stage.

The initial schedule analysis phase. During this phase, the rookie used a medical dispensary as the construction project. The following general characteristics describe the project and the tools to which the rookie had access: a) activity on the arrow; b) 175 activities; c) location was assumed to be Champaign Illinois; d) cost = $600,000.00; e) start = 1 July 1976; f) finish = 14 August 1977; g) PMS = Primavera Project Planner^{Tm}. A total of three hours were allocated to this phase.

The in-progress schedule analysis phase. During this phase, the construction project was a six-story apartment building. The following general characteristics describe the project and the tools to which the rookie had access: a) activity on the node; b) 143 activities; c) location = Champaign Illinois; d) cost was not known; e) start = 1 June 1986; f) finish = 27 October 1987; g) PMS = PMS80^{Tm}. Three hours were assigned to this stage.

The overall schedule analysis phase. This phase was designed to specifically address automatic schedule generation issues as well as schedule logic analysis. No specific project was utilized by the rookie. Rather, the rookie and trio electronically conversed at a conceptual level only. Two hours were devoted to this period.

The de-briefing phase. Because of the uniqueness of the technique's implementation to elicit knowledge, it was paramount to de-brief the trio and rookie. This feedback will be used to modify future attempts to knowledge acquisition. This de-briefing lasted one hour.

Exercise design. In order to keep tight control on the information exchange between the rookie and the trio of experts, the setting consisted of two rooms divided by a partition wall. In one room the rookie was conducting a construction schedule audit using only the reams of reports generated by one or two project management systems and the words of advice coming from the trio of experts. While the video cameras recorded the approach, the experts, who were located in a separate room, guided the rookie in his analysis. The question/answer protocol between the rookie and the trio of experts was implemented in electronic mail messages within a Local Area Network.

The amount of information available to the trio of experts was somehow restricted. The trio had only a tablet and a pencil with which to work. The hypothesis is that if such project information is withheld, we can force the experts to rely heavily upon, and hence provide additional evidence about, their knowledge and reasoning skills [Hoffman 87, Kneale 86]. By looking across the trio of experts' tactics and procedures one should be able to know what data they like to have available and the information that is produced.

For the same test case, an attempt was also made to use the existing rules and procedures, which were derived from the articulation approach and the transformation of public domain knowledge. A fourth expert conducted such assessment and provided the knowledge engineer with the opportunity to simultaneously examine the reasoning of the second generation knowledge base. This comparison led to find points of disagreement and consensus and thus generated a more robust human say-how.

Three points of view were fused during the exercise. They were: a) the contractor, represented by W.E. O'Neil Construction company; b) the owner, portrayed by the U.S. Army Corps of Engineers; and c) the project management software developer, represented by Pinnell Engineering, Inc.

In general, the only difference between a rookie and an expert is the quantity and organization of knowledge. Since a primary objective was to stimulate and elicit knowledge from the trio of experts, the rookie was not as knowledgeable as the experts were. Thus, a junior project manager acted as such. [De La Garza 88] explains the knowledge in plain English elicited during the exercise.

Exercise mechanics. The mechanics of the exercise played an important role in stimulating knowledge. They were:

1. Because of the novelty of the exercise's limited information approach, the trio of experts were at times bothered by the fact that they had to request every single piece of information about which they wanted to know. They would have preferred to have had a narrative description of the project to assimilate as much implicit detail as they could. This would defeat the purpose of using the limited information approach, because the trio would not be cognizant of all the facts and relationships being digested.

2. A more formal question/answer protocol was desired by both the trio and the rookie.

3. The trio should have provided the rookie with just "words of advice" and let the rookie perform the necessary hand-calculations to implement such advice.

4. The trio was frustrated by the amount of hand-calculations they had to perform.

5. The roles of the trio and rookie were dynamic. That is, while the rookie was trying to stimulate and extract as much knowledge as possible from the trio, the trio was trying to obtain relevant information from the rookie. This caused identity confusion a few times.

6. When the trio was given more information than they asked for, i.e., a PMS report containing all critical activities with basic parameters, they spent a considerable amount of time analyzing the report in silence. During two cases, the trio's "thought process" was stimulated but not captured. PMS reports were intentionally given to the trio to point out the disadvantages of using non-limited information approaches to knowledge elicitation.

7. Because the words of advice coming from the trio were not always derived directly from the information supplied by the rookie, the rookie was forced to ask "Why" and "How". The trio would then characterize the rookie as: stupid, obstinate, and too demanding. It is evident that these adjectives reflected the frustration of the trio when they had to think very hard to articulate knowledge.

8. Because the trio was required to think thoroughly and type all questions/answers, the electronic mail was considered slow from a real-time feedback standpoint. This slowness also caused the trio and rookie to be asynchronized.

9. The mechanics of reading/sending/printing messages and attachments were inefficient.

10. The coordination of the "subject:" slot in the template of each electronic message was also difficult. The message replay function of the electronic mail system does not include the body of the message to which it is replaying.

11. The trio began adjusting to the exercise mechanics towards the end of the 2-day exercise.

12. In order to release some frustration built-up by the inability of the trio to converse with the rookie, the electronic communication protocol should allow for a "chat" mode between the trio and the rookie.

13. The lack of a "chat" mode in the communication protocol prevented the trio and rookie from exchanging vague phrases like, "you know what I mean...". This would have made the exercise more realistic in terms of how a person communicates with another person, but not in terms of how he or she interacts with a computer.

14. The time necessary for the trio to build a mental picture of the project was very substantial.

15. The idiosyncrasies of a particular project were not part of the exercise. The trio found much easier to discuss scheduling at a conceptual level than at a specific project level.

16. 3 days' worth of knowledge elicitation time, instead of 3 hours, may be necessary in each of the schedule analysis phases (initial, in-progress).

Exercise consensus. The consensus of the exercise can may be summarized as follows:

1. There is a sufficient common body of construction knowledge that can be meaningfully categorized, structured, and applied within a KBS to make it worth developing it, even to the point of a robust non-computerized set of scheduling guidelines.

   1.a The trio summarized the level of difficulty in assessing the reasonableness of the schedule logic with the statement: "We have a bull by the tail". However, the trio also concluded that the logic aspect can be addressed at a sufficiently generic level and that the burden of structuring and collecting the knowledge will not render the development of a KBS infeasible.

   1.b In the short term a check list approach to determine the reasonableness of the schedule logic for each building system and building type may be enough.

    1.c   The process to evaluate the schedule logic is more qualitative than quantitative.

    1.d   The quantitative aspects of schedule criticism may be generalized for several building systems.

2. The limited information approach proved to be feasible and practical to stimulate and capture knowledge.

3. The trio became aware of how difficult it is to articulate the thought process or the concepts that experts utilize intuitively.

4. Three operation modes for KBS can be implemented:

    4.a   The KBS has access to all project information, performs all necessary calculations, and directly makes inferences based on those.

    4.b   The KBS directs the user to perform hand-calculations whose results are fed back to the KBS. The KBS makes inferences and suggests further analysis.

    4.c   A hybrid approach may be necessary when the procedures have not been implemented in a way that the KBS can execute them, yet, the KBS is capable of reasoning about their results.

5. The user of KBS should be substantially knowledgeable about construction, as opposed to being an apprentice.

6. A KBS for construction scheduling criticism has a high payoff.

7. Even though owners and contractors have slightly different objectives, a single KBS for construction scheduling may be used by both in many circumstances.

8. The knowledge elicitation exercise was interesting, unique, and frustrating. The discussion among the trio of experts was never constrained by the function each expert was representing.

<u>The Knowledge Base Analyzer.</u> The consensus of the hand simulation process is summarized as follows:

1. The performance of the second generation knowledge base was tested during the exercise by comparing its recommendations with those of the trio.

2. The execution of the second generation knowledge base was data driven. That is, all specifications contained within it were expected to react to the PMS data. Therefore, it performed analysis on areas not explicitly covered by the trio, i.e., search for multiple critical paths.

3. The second generation knowledge base addressed a few of the concepts being considered by the trio, i.e., front-end loading, and weather sensitivity.

4. The second generation knowledge base did not addressed many of the concepts covered by the trio, i.e, influence of liquidated damages on the project's criticality.

THE "PAPER" KNOWLEDGE BASE

The "paper" knowledge base consists of a set of schedule provisions written in plain English. These provisions are the result of analyzing and generalizing the knowledge produced by each knowledge acquisition technique. The provisions represent the flesh or the "what" construction schedules need to comply with, and thus, provide no details as to "how" they should be implemented. Two examples within each subcategory are presented next. The complete set of scheduling provisions may be found in [De La Garza 88].

General Requirements.

1) All activities affecting construction progress should be included and defined in a way that they can be easily monitored and measured. The total number of activities should remain manageable.

2) Activities should be described in a way that computers and anyone familiar with the construction work can understand them.

Time.

1) Schedule projections should be based on comparisons between what has happened and what was planned.

2) Float should be broad enough to support the premise that it has not been manipulated.

Cost.

1) Cash flow front-end loading is unlawful.

2) The monetary value associated with an activity should play no role in constraining its duration.

Logic.

1) An activity is weather sensitive if its materials and/or labor are affected by either water, temperature, or moisture.

2) The building enclosure should be logically related to weather sensitive activities.

CONCLUSIONS

A deeper understanding of the process used by owners and contractors to evaluate the goodness and reasonableness of construction schedules has to be achieved. The work described in this paper has contributed with another step in this understanding by articulating good scheduling practice provisions, thus enabling models of the construction schedule evaluation process to be created for subsequent automation.

The three knowledge acquisition techniques were selected based on their own merits and expected results. Their application demonstrated that the research is broadly sound and that there is a sufficient common body of construction knowledge that can be meaningfully categorized, structured, and applied within a KBS to make it worth developing it.

ACKNOWLEDGEMENTS

The authors wish to thank the participation and dedication of the trio of experts in this successful knowledge engineering effort. Mr. Gerald Landsly from W.E. O'Neil Construction Company, Mr. Steve Pinnell from Pinnell Engineering, Inc., and Capt. Scott Prosuch from the Military Science Department, University of California at Davis.

This material is based upon work supported by the National Science Foundation under Grant No. MSM-8451561, Presidential Young Investigator Award, and by the U.S. Army Construction Engineering Research Laboratory under Project No. AT23-AO-048, An Integrated AI-based Project Management System. Any opinions, findings, and conclusions or recommendations expressed in this publication are those of the authors and do not necessarily reflect the views of the sponsors.

REFERENCES

[Avots 85] Avots, I., "Application of Expert Systems Concepts to Schedule Control," Project Management Journal, Vol. 16, No. 1, March, 1985, pp. 51-55.

[Clough 81] Clough, R. H., Construction Contracting, 4th Edition, 1981, John Wiley & Sons.

[De La Garza 88] De La Garza, J. M., "A Knowledge Engineering Approach to the Analysis and Evaluation of Construction Schedules for Vertical Construction," Ph.D. Thesis, Department of Civil Engineering, University of Illinois at Urbana-Champaign, Illinois, March, 1988.

[Feigenbaum 83] Feigenbaum, E. A., and McCorduck, P., The Fifth Generation: Artificial Intelligence and Japan's Challenge to the World, Addison-Wesley, Reading, MA, 1983.

[Freiling 85] Freiling, M., Alexander, J., Messick, S., Rehfuss, S., and Shulman, S., "Starting a Knowledge Engineering Project: A Step-by-Step Approach," The Artificial Intelligence Magazine, Vol. 6, No. 3, 1985, pp. 150-164.

[Gray 86]   Gray, C., "Intelligent Construction Time and Cost Analysis," Construction Management and Economics, Vol. 4, 1986, pp. 135-150.

[Hart 85]   Hart, A., "Knowledge Elicitation: Issues and Methods," Computer-Aided Design, Vol. 17, No. 9, November, 1985, pp. 455-462.

[Hoffman 87] Hoffman, R. R., "The Problem of Extracting the Knowledge of Experts from the Perspective of Experimental Psychology," The Artificial Intelligence Magazine, Vol. 8, No. 2, 1987, pp. 53-67.

[Kneale 86] Kneale, D., "How Coopers & Lybrand Put Expertise into its Computers," The Wall Street Journal, November 14, 1986.

[Michie 86] Michie, D., "Machine Learning and Knowledge Acquisition," in Expert Systems - Automating Knowledge Acquisition- AI Masters Handbook, Edited by Donald Michie and Ivan Bratko, Addison-Wesley Publishers, 1986.

[Mittal 85] Mittal, S., and Dym, C. L., "Knowledge Acquisition from Multiple Experts," The Artificial Intelligence Magazine, Vol. 6, No. 2, 1985, pp. 32-36.

[O'Connor 82] O'Connor, M. J., Colwell, G. E., and Raynolds, R. D., "MX Resident Engineer Networking Guide," Technical Report P-126, U.S. Army Corps of Engineers Construction Engineering Research Laboratory, Champaign, IL, April, 1982.

[Prerau 87] Prerau, D. S., "Knowledge Acquisition in the Development of a Large Expert System," The Artificial Intelligence Magazine, Vol. 8, No. 2, 1987, pp. 43-51.

[Trimble 86] Trimble, G., Bryman, A., and Cullen, J., "Knowledge Acquisition for Expert Systems in Construction," Proceedings of the 10th Triennial Congress of the International Council for Building Research, Studies and Documentation. CIB86 Advancing Building Technology, Vol. 2, September, 1986, Washington, D.C., pp. 770-777.

AN EXPERT SYSTEM FOR CONSTRUCTION CONTRACT CLAIMS

Moonja Park Kim[1] and Kimberley Adams[2]

ABSTRACT: An expert system for claims guidance has been developed at U.S. Army Construction Engineering Research Laboratory. It uses an expert system shell for IBM-compatible microcomputers, Personal Consultant Plus by Texas Instruments. The objectives of this system are (1) to provide an inexperienced project engineer with prelegal assistance in the analysis of potential claims from construction contracts and (2) to serve as a training device for new personnel in field offices, familiarizing them with the related legal issues. This paper describes the development of the first module of the Claims Guidance System (CGS) for analyzing "Differing Site Condition" (DSC) claims. It also includes selection of appropriate cases as basis of the system and a sample consultation.

INTRODUCTION

Among the most significant successes in the field of artificial intelligence has been the development of expert systems. These systems are designed to represent factual knowledge in specific areas of expertise and to provide the problem-solving capabilities equal to those of recognized experts. The potential power of these systems has lead to a worldwide effort to apply this technology to various areas of expertise.

The expert system technology available today on microcomputers makes it possible to address a significant problem facing the construction industry: the need for expertise in construction claim analysis at the field level. Construction in the 1980's has become a very complicated industry, with many intertwined relationships and intense competition. There seems to be a great potential for the applying the state-of-the-art knowledge in expert system technology to the practical areas of construction. One promising area is using an expert system to help minimize some of these problems by providing field personnel some guidance in handling legal issues related to potential claims.

The professionals in the legal field are also taking advantage of this new technology, as evidenced at the First International Conference on Artificial

---

[1] Principal Investigator, U.S. Army - Construction Engineering Research Lab, Champaign, IL 61820

[2] Graduate Student, Dept. of General Engineering, University of Illinois, Champaign, IL 61820

Intelligence and Law held May 27-29, 1987. Some law firms are even creating expert system groups to perform in-house research and development. For example, Watt, Tiedler, Killian and Hoffar, a law firm in Virginia, is developing a microcomputer expert system for claim identification and evaluation (Lester, 1987). The applications of this technology will assist lawyers in sorting out pertinent information for efficient discussion with the client.

Researchers at the U.S. Army Construction Engineering Research Laboratory (USA-CERL) have been developing an expert system called Claims Guidance System (CGS) to provide claims analysis expertise at the field level of the Corps of Engineers. This system uses an expert system shell for IBM-compatible microcomputers, Personal Consultant Plus by Texas Instruments. The objectives of this system are (1) to provide an inexperienced project engineer with prelegal assistance in the analysis of potential claims from construction contracts and (2) to serve as a training device for new personnel in field offices, familiarizing them with the related legal issues.

The first module of the Claims Guidance System (CGS-DSC) will guide project engineers in analyzing "Differing Site Condition" (DSC) claims. Unknown subsurface conditions or latent physical conditions at the work site represent a very significant risk inherent in many construction contracts. The DSC contract clause represents an effort by the U.S. Government to reduce the risk to the construction contractors of such unknown or unanticipated conditions. This clause allows contractors to submit their bids based on reasonably foreseeable conditions, without contingencies to cover the unexpected or unusual. In return, the bidder is assured that in the event conditions prove different than should have been anticipated, an equitable adjustment will be made in the contract price and/or duration. Without this clause, the contractors' only alternative, in order to meet the requirement for submitting a fixed price, was to include contingency allowances in their bids to cover the cost of coping with possible subsurface difficulties, which in fact may not have occurred during subsequent performance of the contract. AS a result, the Government paid more than the actual work was reasonably worth.

Studies by Mogren (1986) and Diekmann & Nelson (1986) have shown that DSC claims are one of the most frequent and costly reasons for changes in U.S. Army Corps of Engineers construction contracts. Corps field engineers who are faced with such claims need to understand the legal issues involved so that they can supply the proper information to legal counsel and avoid lengthy litigations caused by incorrect decisions. Personnel who are unfamiliar with this process must rely on experienced engineers for help in analyzing a claim. The expert system for Claims Guidance is to provide the expertise of the experienced engineers in dealing with construction contract claims.

Specifically, an expert system for analyzing potential claims insures that a rigorous evaluation is performed consistently. It provides a written document of the claim analysis for future reference, which is especially useful if the claim must be defended. In addition, repeated use of the expert system sharpens the field engineers' claims evaluation skills which will help them identify potential claims sooner, avoid conflicts if possible, and support their position with adequate documentation.

This paper describes the expert knowledge acquisition process and the development of the CGS-DSC and a sample consultation session.

ACQUISITION OF EXPERT KNOWLEDGE

Diekmann & Kruppenbacher (1984) have demonstrated that there is significant potential for applying artificial intelligence to claims analysis. They identified the need for more development work in this area to make this technology a viable tool for construction professionals in claims analysis. Following their suggestion and taking advantage of their work on knowledge acquisition, the DSC clause was selected as the first module of CGS for actual use in the Corps field offices.

The expert knowledge acquisition process for the CGS-DSC started with an analysis of the Differing Site Condition Clause (FAR 52.236-2) used by the U.S. Government in its contracts which cites:

(a) The Contractor shall promptly, and before the conditions are disturbed, give a written notice to the Contracting Officer of (1) subsurface or latent physical conditions at the site which differ materially from those indicated in the contract or (2) unknown physical conditions at the site, of an unusual nature, which differ materially from those ordinarily encountered and generally recognized as inherent in work of the character provided for in the contract.

(b) The Contracting Officer shall investigate the site conditions promptly after receiving the notice. If the conditions do materially so differ and cause an increase or decrease in the Contractor's cost of, or the time required for, performing any part of the work under this contract, whether or not changed as a result of the conditions, an equitable adjustment shall be made under this clause and the contract modified in writing accordingly.

(c) No request by the Contractor for equitable adjustment to the contract under this clause shall be allowed, unless the Contractor has given written notice required; provided, that the time prescribed in (a) above for giving written notice may be extended by the Contracting Officer.

(d) No request by the Contractor for an equitable adjustment to the contract for differing site conditions shall be allowed if made after final payment under this contract.

From the above clause, the following important issues in dealing with DSC claims are identified:

1. Final payment: A Contractor that has accepted the final payment is not allowed to file a claim, as described in (d) above.

2. Notice Requirements: The Contractor must give proper notice of the differing site condition in order to maintain the possibility of entitlement as described in (a) and (c) above. Key aspects here are promptness, the receipt of the claim notice by the responsible Government personnel, and the written form of the notice.

3. Prejudice to Government: If a failure to give written notice has not resulted in any prejudice to the Government, the Contractor's right to relief on a valid claim will not be barred. If the Government's interests

have been impaired by the Contractor's failure to give proper notice of the differing site conditions, it is unlikely that the Contractor will be entitled to compensation. To check if the Government was prejudiced, itis necessary to check if the Government would have directed the same actions had it received the appropriate notice.

4. Government Action: The contracting officer must look into the problem of the site condition as soon as he receives the notice in order to discuss the problem with the contractor and to direct the actions to be taken by the contractor for the problem, as described in (b) above.

5. Contract Provision (Type I or Type II): The DSC clause gives two avenues of recovery for the contractor, depending on the contract provisions that exist in each particular case. A Type I case is characterized in (1) of the above clause. The contractor can be entitled to compensation if the actual conditions differ materially from the conditions explicitly or implicitly mentioned in the Contract. A Type II case is characterized in (2) of the above clause. The contractor can also be entitled to a compensation if there is no specific indication of the condition in the contract documents, and if it can be demonstrated that by following the standard construction practice the Contractor could not expect, nor detect, the presence of differing site conditions prior to bid time.

6. Acceptable and prudent: When there was no indication of condition in the contract document, the contractor can make assumptions that are generally accepted practice in the construction industry. These assumptions must be same as what a prudent contractor normally would use (assume usual and known conditions).

In addition to the analysis of the DSC clause, the work of Diekmann and Kruppenbacher (1984) was considered in the process of knowledge acquisition. The logic diagram of Kruppenbacher's (1984) study was reviewed by an experienced Corps field engineer and was revised and simplified to fit the Corps office environment. The questions used in Kruppenbacher's system include many legal terms that could confuse the field engineers; therefore, questions for the CGS-DSC were changed to be easily understandable by the Corps field office personnel. Using the revised logic diagram and questions, rules were developed to create a test version of the CGS-DSC.

A steering committee was formed to review the test version and to evaluate it for validity and completeness. The committee consisted of six experts: two experienced legal counsels from the Corps headquarters and four engineers with many years of experience in construction contract management within the Corps. The committee suggested many enhancements and necessary corrections to the logic diagram and identified the following additional issues as important in handling DSC claims :

7. reliance
8. superior knowledge
9. nature of condition (Act of God)
10. site inspection.
11. material difference
12. exculpatory language
13. anticipation: usual and known

Two legal case retrieval systems, Lexis[3] and Weslaw[4], were used to select appropriate cases which represent some of the issues listed above. Twenty-three relevant cases were retrieved. These cases were examined carefully in terms of the 13 issues and of the rationale for the decision of the Board of Contract Appeals and/or the Court of Claims.

Short descriptions of four of these cases are included here to explain how different cases represented different issues in the CGS-DSC. The first case, C.H. Leavell v. Eng BCA (No.3492, 1975), added the issue of reliance. In this case the Contractor's claim for a DSC equitable adjustment was denied because the Contractor mistakenly relied on inconsistent contract information when more detailed information was available. The Contractor would have discovered the information if he had followed the leads in the contract itself. The Contractor also failed to make his DSC claim until one year after the work was completed. Since the Contractor failed to meet the notice requirement, the Government was prejudiced and could not defend the case adequately. This contract was to construct five buildings at Lackland Air Force Base. The contract's borings indicated that the subsurface soil was practically impervious. Some of the drawing's symbols were unclear but the Contractor assumed that the symbols represented impervious soils and that water would not enter the holes after drilling. Unfortunately, the soil was pervious and extensive casing was necessary for many of the drilled piers.

The second case, the Portable Rock v. U.S. (Court of Claims, 1984), added the issue of superior knowledge. The Contractor sought an equitable adjustment for a Type I DSC condition when he found subsurface water conditions, not specifically mentioned in the contract, which substantially increased the costs of performing the contract. The Contractor alleged that the Government had knowledge of this condition but did not reveal this information to the bidders. The Contractor's claim for an equitable adjustment was denied. The Court held that the Government did not withhold information from the Contractor. Furthermore, the contract interpretation must be reasonable, and it was not in the case. The requirements in the contract, such as the reinforced subgrade, were a clear indication of unstable soil conditions. This should have alerted a knowledgeable contractor to the presence of ground water and wet ground conditions. This Contractor had superior knowledge because he had encountered wet conditions in this area when constructing another road, so he was aware of the conditions.

The third case, Turnkey Enterprises, Inc. v. U.S. (Court of Claims 1979), added the issue of "Act of God" to CGS. The contract was for the repair of seven damaged sites on the Mad River Road in the Six Rivers National Forest in northern California. The Contractor claimed both Type I and Type II DSC conditions when a drought caused his costs to increase. The Court denied both Type I and Type II claim for an equitable adjustment, because a drought is an "Act of God" and not any fault of the Government. And also the contract documents do not indicate anything about the alleged changed condition. There was nothing in the contract on which the Contractor could claim he relied upon. There was no guarantee that water would be in the

---

[3] Lexis is a Trademark of Head Data Central, Inc.

[4] Westlaw is a trademark of West Publishing Company.

river at all times. Generally, the Government does not assume an obligation to compensate a contractor for additional costs or losses incurred as a result of solely weather conditions.

The fourth case, Parker Construction Co. v. U.S. (Court of Claims 1970), added the issue of reasonable site inspection. The Contractor entered into a unit price contract to build 2.366 miles of graded road in Washington State Mt. Baker National Forest. The Contractor claimed a Type II DSC because of hard rock encountered that increased his drilling costs. The Government claimed that the Contractor did not conduct a prudent site inspection. The Contractor assumed that the brownish cleavage faces of the rock indicated that the rock was softer and easier to drill than solid gray rock. He assumed this even though there were outcrops of hard rock apparent upon visual examination. Furthermore, the Contractor failed to prove his Type II DSC claim. The Contractor had to prove that he encountered something materially different from the "known" and "usual." The Court denied the Contractor's claim for an equitable adjustment.

DEVELOPMENT OF THE CGS-DSC

The logic diagram for the CGS-DSC was developed based on important issues identified from the analysis of cases in combination with the logic diagram of Kruppenbacher (1984). The logic diagram and listing of 23 retrieved case briefs are found in the reference manual of CGS.

Based on the logic diagram, rules for the CGS-DSC knowledge-base were written using Personal Consultant Plus. There are about 300 rules and 95 questions that require the user's input in this knowledge base. The user is not asked all of 95 questions in every consultation; only a portion of them will be asked depending on the answers the user enters.

In selecting an expert system shell for the development of the system there were some limitations: (1) Consultation should be available on IBM-compatible computer with 640K memory, (2) Cost of developing and delivering system should be minimized. Personal Consultant Plus was selected to meet these limitations and to provide a user friendly interface.

As we added rules to the knowledge-base, it was necessary to add 1.5 megabytes of memory for the development stage. However, every effort was made to make the delivery system run with 640K memory, by creating disk-resident text files for help screens and case summaries. As shown on Figure 1 on the next page, the CGS-DSC delivery system environment was created by displaying text files for help screens, by presenting PC Paint graphic images for progress checks, and by running dBASE III Plus to search for the appropriate case name of the text file to be displayed. In order to create a useful report for the users, dBASE III Plus was used to organize input entered by the user and to provide a documentation of facts. An example of a help screen and a case summary are included for the sample consultation discussed below.

The field test version was delivered to two test sites with proper training in November 1987. A one-year the field test is planned; however, suggestions from the users will be incorporated as they are received. Then the final system will be distributed to a number of field offices for regular uses.

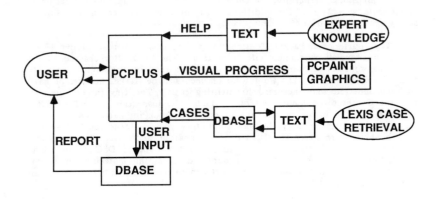

Figure 1: CGS-DSC Environment

A SAMPLE CONSULTATION

One case from the 23 retrieved cases was chosen to be used as an example for explaining the use of the CGS-DSC: Hallmark Electrical Contractor Inc., v. ASBCA No. 32595 (April 27, 1987). A short description of this case is as follows:

> The Contractor was awarded a contract to replace an overhead electrical distribution system at the Belton Lake Recreational Area near Fort Hood, Texas. Contractor was required to dig holes approximately six feet deep and place wooden poles in them. The Contractor estimated that it would take about an hour to dig each hole because he had worked on similar terrain about ten miles away from the present site at Fort Hood. Unfortunately, the Contractor encountered problems in digging and could not use the standard combination dirt/rock bit. The Contractor encountered "blue rock" which he had never encountered and which was much harder than the more prevalent "butter rock." The Contractor submitted a claim under both Type I and Type II DSC. The Contractor was denied recovery under Type II DSC because the condition encountered did not differ materially from the "usual" and "known." Contractor was not entitled to recover under Type I DSC either, because the contract did not indicate conditions different from what the Contractor encountered during performance.
>
> A Final Payment release was not signed by the Contractor. The Contractor complied fully with the notice requirement by filing a timely claim with the Contracting Officer for the adjustment. It was denied in its entirety by the Contracting Officer.

The contractor anticipated butter rock and encountered blue rock and argued that blue rock is unusual for the area where he worked. However, an expert witness testified that "blue rock" is usual and known for the area. Therefore, the contractor was not prudent to anticipate butter rock when submitting his bid.

The contractor did not conduct a reasonable site inspection because he assumed that the absence of a "rock table" is an "indication" that rock would not be encountered. It was apparent during a site visit because there were rock outcroppings all over the site, but he did not take this into consideration in his bid.

<u>Rationale For Denying An Equitable Adjustment:</u> The Contractor failed to prove that he encountered something materially different from the "usual." It was true that the blue rock was significantly harder than the butter rock. This in itself did not establish a Type II DSC. The Contractor had an expert testify, but the expert did not establish that the hard material was geologically unusual in the area. In fact, he established just the opposite. Since the Contractor encountered material within the range of hardness usually encountered, the Contractor was denied his Type II DSC.

This case will be used for demonstrating a sample consultation of the CGS-DSC. The user starts the consultation by typing "CONSULT CLAIMS". Then the title and the objectives screen will be displayed. The objective screen is shown below:

```
            CLAIMS GUIDANCE SYSTEM
       This Claims Guidance System is developed
       to provide:

       1) basic awareness of the issues
          surrounding the DSC clause,
       2) measuring device to ascertain
          contractor's chance of DSC entitlement,
       3) documentation in the event the dispute
          comes to trial.
     * End - RETURN/ENTER to continue
```

Next, the user is asked to type in relevant information for documentation, such as the name of the contractor, contract number, so on.

    Q: Please enter the name of the contractor.....
                <u>Hallmark Electrical Contractor's, Inc.</u>
    Q: Please enter the contract number.....
                <u>DAKF48-85-C-0052</u>
    Q: Please provide the description of the contract.....
                <u>Overhead Electrical Distribution System</u>
    Q: Please enter your name (the user).....
                <u>John Smith</u>
    Q: Please give a brief statement of the contractor's assertion.....
                <u>The contractor could not dig holes using the standard combination dirt/rock bit in the one hour anticipated. It took as much as a day or more to dig each hole where</u>

'blue rock' was encountered. It took about 45 days to drill all the holes rather than the ten days the contractor had anticipated.

For all questions displayed below, the answers are shown in the parenthesis ( ). However, in actual consultation, the user will move the cursor to the selection.

Q: Has the contractor signed the final payment release without condition ?.................... **(NO)**
Q: Was the final payment sent to the contractor ? .... **(NO)**

The next step is to find if the contractor had complied fully with notice requirements. In order to check if appropriate notice was given, the following questions will be asked:

Q: Did the contractor inform the government about the problem prior to asserting a differing site condition ? ...... **(YES)**
Q: How did the contractor inform the Corps of the problem?........ .... **(WRITTEN)**
Q: Please enter the name of the person that received the information first.....
   Mr. James Kemp
Q: Please enter the date the information was received.....
   JUNE 27, 1985
Q: Who was the first person or group to receive information regarding the claim ? ... **(RCO/COR OR AUTHORIZED REPRESENTATIVE)**

At any point before answering a question, the user can invoke the WHY option requesting to explain why the system is prompting for this information. The why screen for the above question is shown as an example:

```
WHY

Who was the first person to receive the
notice is needed to determine if responsible
person did receive the notice.

RULE 008
In order to check if the contractor complied
fully with the requirements for the notice
the following information is necessary:

(1)  if responsible personnel received the
     notice and the date received,
(2)  if the form of notice was adequate
     (written or oral)
(3)  if the notice was prompt.

     ** End - RETURN/ENTER to continue
```

Q: When did the contractor first inform the government about the problem ? .... **(BEFORE STARTING WORK ON PROBLEM)**

From the answers shown above, a conclusion can be made that the contractor complied fully with notice requirements by informing the responsible personnel in a written form before starting work on the problem. Now we need to check if the government's action was adequate with the following questions:

> Q: Did the government investigate the asserted Differing Site Condition ? .... (YES)
> Q: Did the government's investigation confirm the contractor's asserted DSC ? .... (NO)
> Q: Please summarize the investigation made by the Government...
> <u>Government decided not to presurvey and represent the subsurface conditions. The contractor must demonstrate that he has encountered something materially different from the 'known' and 'usual'. Even though the government found that blue rock was harder than the softer material that the contractor hoped for, this is not unusual around this area.</u>
> Q: Did the government provide the contractor with directions concerning the asserted DSC ? .... (NO)

At various points a graphic display of progress through the system is available. An example of this graphic display after checking the Government action is shown in Figure 2. This graphic display is an over-simplified version of the CGS-DSC logic diagram. It informs the user how many of the issues were covered through the system at various stages.

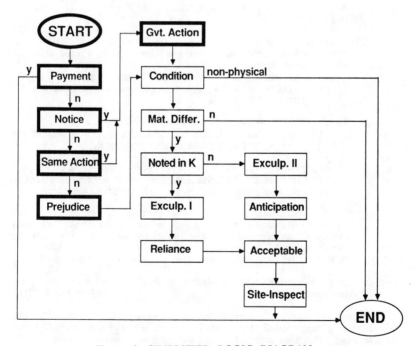

Figure 2: SIMPLIFIED LOGIC DIAGRAM

The next step in the CGS is to check the nature of the problem, material difference, contract provisions, usual and known anticipation issues.

Q: Is the asserted DSC directly related to the physical conditions at the site ?...................... **(YES)**

Q: Did the asserted DSC exist before the contract award ? **(YES)**

Q: Does the physical condition substantially differ from the indicated/anticipated condition ?.......... **(PROBABLY YES)**

Q: Please describe why you believe that the condition differs substantially...
<u>Because the blue rocks the contractor encountered are much harder than what the contractor hoped to encounter, and it took much longer to drill holes which resulted in substantial increase of cost.</u>

Q: Did the asserted DSC increase the contractor's costs/duration ? ... . **(YES)**

Q: Did the contractor utilize acceptable practices or standards of workmanship to alleviate the DSC ? .... **(YES)**

Q: Are there contract indications or notes on drawings that pertain to this situation ?................... **(NO)**

Q: Please explain in detail why you believe that the contract does not either impliedly or explicitly make reference to the alleged differing site condition claim.....
<u>The government has elected not to presurvey and represent the subsurface conditions, so there was no indication about the condition in the contract.</u>

Q: Please explain the known and usual conditions in the area.....
<u>Blue rocks can be found scattered around this area and the hardness of these blue rocks varies a great deal.</u>

Q: Please explain the conditions that were encountered in the area......
<u>In one third of the holes, hard blue rocks were encountered so that the contractor spent a day or more to drill each hole.</u>

Q: Is there a clause that limits the government responsibility and liability by stating that indications on the contract are only representation of conditions and should not be a basis for differing site condition ?................... **(YES)**

Q: Is this clause extremely specific to the condition encountered ?. . . .. **(NO)**

Q: Did the contractor act prudently in making assumptions about the site condition ?............. **(DEFINITELY NO)**

At any time during the consultation before answering a question, the user can display a help screen for explanation of the question, as additional clarification. The help screen for the last question above is shown on the next page:

> **PRUDENT CONTRACTOR:**
>
> If most contractors with similar experience and knowledge, and under the same circumstances, would agree with the contractor in question, the contractor is said to be a prudent one.
>
> Prudent Contractor would:
> (1) Utilize his past experience and knowledge as a basis for all his judgments and assumptions.
> (2) Act as most of his peers would given the same set of circumstances.
> (3) Submit contract bid based on fair reading of representations indicated in government contract and adequate site inspection.
> (4) Make assumptions utilizing his past experience and knowledge when conducting a site inspection.
> (5) Make assumptions utilizing his past experience and knowledge when reading the contract drawings, specs., and, therefore, not rely on obvious flaws noted in the contract.

Q: Please explain why you think that the contractor was not prudent....

<u>He should have checked what is usual around this area, and he should have noticed from the site visit of rock outcroppings all over the site.</u>

Q: Should the contractor have known about the DSC prior to bidding ?.
.... **(DEFINITELY YES)**

Q: Is there evidence that the contractor knew of the DSC prior to bid ?.......................... **(NO)**

Q: Did the government have current knowledge of the DSC prior to bid ?.......................... **(NO)**

Q: Would a reasonable site inspection have provided information to detect the DSC ?................... **(YES)**

Q: Did the contractor conduct a reasonable site inspection prior to bid ?..................... **(DEFINITELY NO)**

Q: Did any act of the government hinder the site inspection ? .....
.... **(NO)**

This is the end of consultation for this case. Next, the conclusion screen will be displayed as shown below:

CONCLUSIONS

Possible conclusions for entitlement are the following:
1. VERY-POOR-CHANCE
2. POOR-CHANCE
3. DIFFICULT-TO-DECIDE
4. FAIR-CHANCE
5. GOOD-CHANCE
6. EXCELLENT-CHANCE

The chance of entitlement for this contractor is as follows: VERY-POOR-CHANCE

After displaying the conclusion reached by the CGS, the system will search for a case similar to the situation described by the user. It may take a few minutes for the system to perform the dBASE III commands and to find the appropriate case. The user will see the dBASE III logo and license agreement statement, then the short summary of the retrieved case from cases database would be displayed. For this consultation, a one-screen summary of the Hallmark case will be displayed.

Next, the list of the pertinent factors in reaching the conclusion is displayed on the screen as shown below:

The list of pertinent factors in reaching the conclusion is as follows:

CONTRACTOR EXPECTATIONS ISSUE
SITE-INSPECTION ISSUE

After this, the user needs to save all his answers in a file in order to review it later. The answers and information which the user has supplied during the consultation can be compiled into a report so that the information can be easily and clearly presented in a fact sheet. This report generation is performed by dBASE III because PC Plus does not provide adequate reports. A sample report is included in the CGS users manual.

FUTURE RESEARCH

Even though USA-CERL attempted to avoid legal terminology in writing questions for users (mainly engineers) the CGS still requires some legal judgment as input. For example, the user has to make a legal judgment to characterize the difference as "material" or "substantial" when answering the question "Does the physical condition substantially differ from the indicated condition?" It would be desirable to include more expert knowledge of lawyers on how to make a legal judgment on "materiality" and other topics. It seems that work on this area may be available in the near future if we look into the research and development being done in law profession.

Many law professionals are actively exploring expert system technology, as evidenced at the first International Conference on Artificial Intelligence and Law held May 27-29, 1987. Waterman, Paul and Peterson (1986) reviewed existing expert systems for legal decision-making potential and indicated that they expect more applications in the following areas: organizing case information, estimating case value and establishing strategies for negotiation, monitoring legal data bases to find changes in the law, interpreting the law in the context of a problem, and producing legal documents.

Another approach was presented in Victor's (1984) article, "How Much is a Case Worth," which demonstrated how a collection of decision trees, subjective probability assessments and calculations can be used in evaluating claims; this application helps trial lawyers assess the monetary worth of alternative courses of action.

Thus, it seems that a great amount of research and development is expected in the near future. We may take advantage of this interest in the legal field and include more legal expertise in improving the CGS to include other types of construction contract claims. For example, including construction delay claims would involve integrating scheduling and network analysis with legal evaluation of claims. Design deficiency claims would involve integrating CADD systems with claims evaluation to examine drawings for deficiencies. These areas are challenging and hold potential benefit for the construction community.

ACKNOWLEDGEMENT

The authors would like to express their gratitude to Vida Florez, student at Vermont Law School, South Royalton, VT 05065, for retrieving and reviewing appropriate cases for this paper.

REFERENCES

Contract Clauses - Construction - Inside the U.S. (1986). Corps of Engineers, Department of the Army.

Diekmann, J. E., and Kruppenbacher, T. A. (1984) "Claims Analysis and Computer Reasoning." Journal of Construction Engineering and Management, Vol. 110, No. 4, **ASCE**, 391-408.

Diekmann, J. E., and Nelson, M. C. (1985) "Construction Claims: Frequencies and Severity." Journal of Construction Engineering and Management, Vol. 111, No. 1, **ASCE**, 74-81.

Kruppenbacher, T. A. (1984) The Application of Artificial Intelligence to Contract Management. USA-CERL Technical Manuscript P-166 (U.S. Army Construction Engineering Research Laboratory).

Lester, J. L. (1987) "Lawyer on a Microchip." Civil Engineering Magazine, ASCE, 68-69.

Mogren E. T. (1986) "The Causes and Costs of Modification to Military Construction." Master Thesis, U.S. Army Command and General Staff College.

Victor, M. B. (1984) "How much is a Case Worth ? - Putting Your Intuition to Work to Evaluate the Unique Lawsuit." Trial.

Waterman, D. A., Paul, J., and Peterson, M. (1986) "Expert Systems for Legal Decision Making." Expert Systems - The International Journal of Knowledge Engineering, Vol. 3, No. 4, Learned Information Ltd, Oxford, 212-226.

KNOWLEDGE ENGINEERING IN A KNOWLEDGE-BASED SYSTEM FOR
CONTRACTOR PREQUALIFICATION

Jeffrey S. Russell,[1] S.M. ASCE and
Miroslaw J. Skibniewski,[2] M. ASCE

ABSTRACT

This paper presents the knowledge acquisition strategy employed in the development of a contractor prequalification knowledge-based system. A discussion of integrating analysis and heuristic tools to aid in decision-making is given. An example application regarding a selected part of the knowledge engineering effort is presented.

1. THE NEED FOR CONTRACTOR PREQUALIFICATION

From the inception of a construction project, an owner must make numerous decisions which result in the success or failure of the entire project. The task of selecting the "right" bidders for a particular project is one such decision. It is one of the most challenging tasks performed by an owner or contract administrator. Every construction project faces adversity and uncertainty which must be overcome. No matter how meticulous the development of the contract, a poor selection of contractor(s) to execute the work will only magnify the problems encountered on the project. Therefore, it is of paramount importance that prior to the bidding process, contractor prequalification should be performed. A prequalified contractor should be competent and able to execute the assigned project.

Prequalification is a process of determining a candidate's competence or ability to meet the specific requirements for a task. In the construction industry, prequalifying a contractor involves the screening, by a project owner, according to a given set of criteria to determine each contractor's competency to participate in a bid. Contractor prequalification is a decision-making process which can be characterized as consisting of a wide range of criteria for which information is often qualitative and subjective. The process remains largely

---

[1]Graduate Research Assistant, School of Civil Engrg., Purdue Univ., W. Lafayette, IN 47907; V.P., Russell Construction, 2055 Watson Ave., Alliance, OH 44601.
[2]Assistant Professor, School of Civil Engrg., Purdue Univ., W. Lafayette, IN 47907.

an art, where subjective judgment, based on the individual's experience, becomes an essential part of the process. Previous discussion of selected topics regarding contractor prequalification include a discussion of bidder selection by (Diekmann 1981) and tender evaluation by (Nguyen 1985).

The process of contractor prequalification has received little research attention, even though it is one of the first steps that owners must take to ensure the eventual success of a construction project. The following reasons for this lack of research attention can be identified (Russell & Skibniewski 1987b):

- Some owners have relied on surety companies providing bonds for contractors for prequalification. Therefore, many owners have not developed a well-structured, formalized prequalification process.

- Some owners do not feel that prequalifying contractors is important enough to warrant the expenditure. However, owners may be subjecting themselves to the risk of admitting contractors with inadequate ability, capacity and experience into the bidding/negotiation process.

- The cost which owners' incur in developing, implementing, and evaluating the prequalification criteria and the evaluation of the contractors.

Numerous benefits can be realized by all parties involved in the construction process, especially owners, contractors and surety companies by performing prequalification. The following reasons, among others, for these benefits can be identified:

FOR OWNERS:

- Efficient execution of their projects by the efficient application of a contractor's resources to the project. This can be measured by construction performance parameters such as time, economic cost, quality and safety.

FOR CONTRACTORS:

- Optimize the rate of return on their money by not expending resources on projects they are less capable of performing efficiently.

FOR SURETY COMPANIES:

- Maximizing own profits by minimizing the risk of contractor failure to execute the project.

This paper presents a structured review of several steps and issues regarding the development of a knowledge-based system for contractor prequalification. In the subsequent section, the knowledge acquisition strategy utilized in the development of the system is outlined. Next, a discussion on integrating analysis and heuristic tools to aid in the decision-making process is given. An example application regarding a selected part of the knowledge engineering effort is presented.

2. KNOWLEDGE ACQUISITION STRATEGY

Before the knowledge base for the contractor prequalification knowledge-based system is designed, the compilation of task-specific knowledge including formalization (identification, definition and analysis) and structure of the prequalification decision support knowledge is required. A significant amount of effort has been expended on this activity (Russell & Skibniewski 1987c), (Russell et al. 1987d). This is due to the numerous iterations and refinements performed in formalizing the knowledge.

The major thrust of the knowledge acquisition activity is to formalize the knowledge used by an expert(s) in contractor prequalification decision-making. Numerous techniques or methods to perform this activity can be applied. A flow diagram representing the approach to knowledge acquisition in regard to the analyzed domain is presented in Figure 1. The process consists of three distinctive steps which are described in more detail below.

In the first step, general interviews were conducted to explore the prequalification process as it is currently performed through 8 personal and 24 telephone interviews. Each individual participating in the study was asked to describe their prequalification process (i.e. describing the domain problem by discussion (Waterman 1986)). Pertinent questions asked related to their description included:

1. Characteristics of the decision maker or owner (e.g. type of construction activity, contracting strategy);

2. Contractor data sources (e.g. past performance data files, questionnaire, Dun and Bradstreet—a credit rating service company report);

3. Decision parameters utilized in the process (e.g. financial stability, experience);

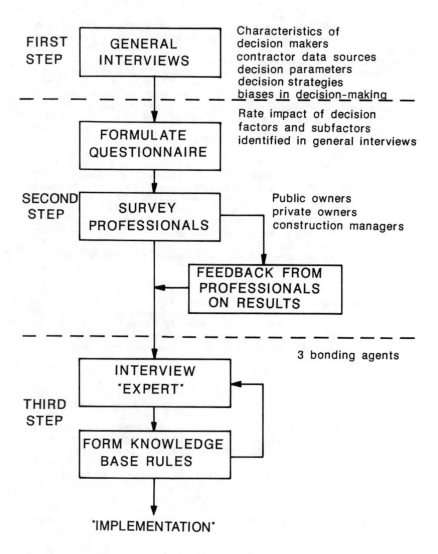

FIGURE 1. Knowledge Acquisition Process

4. Strategies employed in decision-making process (e.g. formula, subjective judgment);

5. Biases in the decision-making process (e.g. owner construction experience, resource constraints).

The study participants included private (15) and public (6) owners, construction managers (5), architects (3) and surety companies (3). The characteristics of the study participants varied by size, location and type of construction activity. The majority of the interviews were conducted with owners and construction managers of large firms (annual construction volume over $25 million) located in the Midwest. Thirteen of the study participants engaged in general building construction, thirteen in industrial construction and three in heavy and highway construction.

A summary of these findings includes:

1. A generic prequalification logic is used in the selection decision regarding the group of contractors admitted to a bid;

2. The prequalification process can be interpreted as consisting of two inputs and one output;

   a. <u>Decision Maker (input)</u>: represents characteristics of the owner that impact the selection of the decision strategy and the formulation of prequalification criteria;

   b. <u>Contractor (input)</u> represents characteristics of the contractor being prequalified;

   c. <u>Decision (output)</u>: utilizes the information provided from the two previous steps to make the prequalification decision.

3. There is a variety of decision factors utilized by the contacted individuals which are characterized and measured based on diverse criteria;

4. Several alternative decision strategies are known to exist;

5. There are numerous factors that can introduce biases into an owner's decision.

A more complete description of these results are given in (Russell & Skibniewski 1987c). These findings impacted the subsequent step in the knowledge acquisition process which is discussed below.

The second step was the development of a questionnaire. The decision to use a questionnaire resulted from the small number of interviews (32) and from the lack of structure to the decision parameters utilized by the interviewees in their prequalification process. Therefore, the questionnaire presented in (Russell et al. 1987d) was developed to gather structured data regarding the pertinent decision factors of the process.

Figure 2 presents an excerpt of the questionnaire format. The questionnaire was structured to facilitate the development of a hierarchical tree representation of all decision factors. An example of this representation is presented in Figure 3. As shown, the decision factor labeled "financial stability" can be characterized by four subfactors labeled "credit rating," "banking arrangement," "bonding capacity" and "financial statement."

The questionnaire incorporated 20 major decision factors (e.g. financial stability, management capabilities, etc.) and 67 decision subfactors that were identified in the general interviews. A decision subfactor is defined as an aid in making a part of a decision associated with a major factor. As shown in Figure 2, the decision factors and subfactors were presented so that an individual could describe how much each decision factor and subfactor impacts their decision process on a scale from 0 to 4. The response alternatives ranged from 0 (little or no impact) to 2 (moderate impact) to 4 (high impact). This particular step in the knowledge acquisition process can be referred to as dissecting the domain by dividing goals (Hart 1986, Grover 1983).

Three organization types participating in the study were public owners, private owners and construction managers. A sample of 344 construction professionals was compiled from various listings of professional organizations and from professional contacts of the authors. A total of 192 questionnaires were returned which represents an overall response rate of 56%.

These data were evaluated by qualitative and quantitative analysis techniques. The qualitative analysis technique involved rank ordering the mean impact of all the questionnaire items for each owner group. The 10 questionnaire items with the largest and smallest mean impact were analyzed. The quantitative analysis techniques included two statistical tests.

CONTRACTOR REQUALIFICATION 175

The following factors deal with prequalifying a contractor. What impact does each of the factors have on the successful prequalification of a contractor?

FACTOR	HIGH		IMPACT MODERATE		LITTLE/ NONE
a. Financial Stability	4	3	2	1	0
b. Experience	4	3	2	1	0
c. References	4	3	2	1	0
d. Staff Available	4	3	2	1	0

There are several subfactors that are used to determine the condition of the major factors. What is the impact of subfactors listed below and others not listed here on determining the condition of a major factor?

EXPERIENCE

Success of completed projects	4	3	2	1	0
Size of completed projects	4	3	2	1	0
Number of similar completed projects	4	3	2	1	0
Types of projects completed	4	3	2	1	0
Other _____	4	3	2	1	0

FIGURE 2. Excerpt of Questionnaire Format Sent to Prequalification Organizations

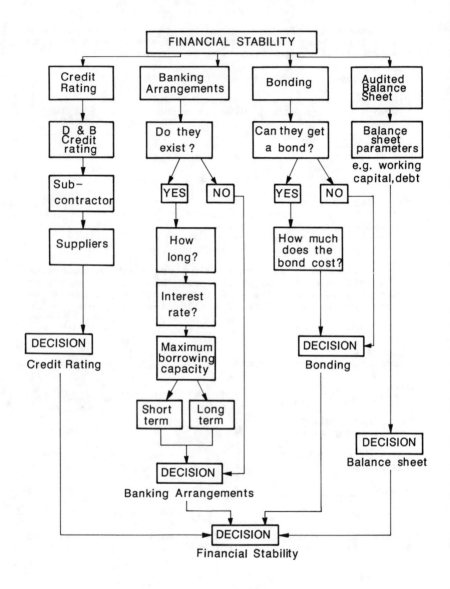

FIGURE 3. Characterization of the "Financial Stability" Decision Factor

The Kruskal-Wallis test was performed on the grouped data. Details of this test method are described in (Conover 1980). This test was performed to compare each questionnaire item's mean impact response across the various organization types to determine if the mean impacts are statistically different at the 0.05 level of significance. The second test, Factor Analysis, was performed on the 20 major decision factors using principle component factor analysis with varimax rotation. A discussion of this technique is described in (Harman 1976). The purpose of performing factor analysis was to identify a reduced number of distinctive composite decision factors relevant in the decision-making.

The major findings of this analysis include:

1. Public owners view the impact of the decision parameters significantly different than private owners or construction managers, while private owners and construction managers view them similarly;

2. Seven distinctive decision factors for public owners and nine composite decision factors for the private owners and construction managers are relevant in the decision-making. These factors are presented below, according to each owner group in Table 1.

GROUP	
Public Owners (1)	Private Owners and Construction Managers (2)

Performance
Capacity for Assuming
  New Projects
Location
Financial Stability
Type of Contractor
Percentage of Work
  Performed
Third Party Evaluation
	Performance
	Capacity for Assuming
	  New Projects
	Location
	Financial Stability
	  and Experience
	Management
	Safety
	Resources
	Failed Performance
	Bonding Capacity

TABLE 1. Major Factors Utilized in the Contractor Prequalification Process

A more complete description of these results can be found in (Russell et al. 1987d).

The third step, which is currently being performed, consists of interviewing construction professionals ("experts") to develop qualitative and quantitative rules. The criteria used for the selection of "experts" for this project was based on performance and experience in the construction industry. In order to identify an "expert," his performance level was measured by the number of failed contracts with which they have been associated. Construction experience of at least 15 years was required. The characteristics of an "expert" are:

1. Bonding agent - 20 years of experience in the bonding industry located in the East Coast and Midwest areas, 1 defaulted contractor;

2. Bonding agent - 27 years of experience in the bonding industry located in the Midwest area, 1 defaulted contractor;

3. Bonding agent - 30 years of experience in the bonding industry located in the Midwest area, 2 defaulted contractors.

The third step consists of two parts:

1. Interviewing experts by describing rules of general knowledge made up by the knowledge engineer and asking them for comments.

2. Procedural simulation or problem analysis to present the experts with actual prequalification cases. A discussion of this technique is given by (Waterman 1986) and (Grover 1983).

An example of part one in the third step of the knowledge acquisition process is described below:

Knowledge Engineer: If the estimated cost of the project the contractor is seeking prequalification on is three times as large (dollar volume) as any project the contractor has completed previously, would you prequalify this contractor?

Expert: The contractor would probably not be prequalified. However, I would want to know why the contractor thinks that he is able to perform the project. Also, I would be interested in knowing how large (dollar volume) their current work program is.

Knowledge Engineer: If the cost of the project bond is greater than 5% of the estimated project cost, what would you conclude about this contractor?

Expert: The contractor is perceived by the bonding company as constituting a high risk. I would ask the bonding company why they perceive the contractor to constitute such a high risk.

Knowledge Engineer: If a contractor has a debt to net worth ratio greater than three to one and working capital is less than $0.00, what would you conclude about this contractor?

Expert: I would draw two conclusions: first the contractor is highly leveraged and is not carrying a majority of the financial risk, and second the contractor is experiencing cash flow difficulties and the contractor's banking arrangements should be checked.

Part two in the third step of the knowledge acquisition process involves a case study underway including construction of a large multi-story office complex in the Midwestern region of the United States. A lump sum bid was requested from the contractors considering this project. Approximately 10 contractors are competing for admission to the bid. Each expert is given pertinent information regarding the project specific characteristics and characteristics of each candidate contractor. The experts provide a rationale for selecting or rejecting each candidate contractor. This rationale provided will aid in extracting and developing qualitative and quantitative rules regarding contractor prequalification for performing general building projects.

3. INTERFACING ANALYTICAL AND HEURISTIC TOOLS

Each application domain consists of diverse characteristics and peculiarities which need to be considered before applying a knowledge-base system or an expert system to a problem solution. One issue facing the knowledge engineer in the development process is which analytical tools (computer simulation, statistics, and others) should be fused with the heuristic tools. Some application domains require the manipulation of large quantities of data in the decision-making while others require very little. It can thus be concluded that the decision how to integrate analytical tools with heuristics and in what proportions is domain-dependent. It is up to the judgment of the knowledge engineer to determine what are the characteristics of the domain problem and to establish a proper proportion and match between the available analytical and heuristic tools (e.g. search strategies: forward or backward chaining; knowledge representation: rules, frames, semantic networks).

An example application of several tools to the contractor prequalification domain is described below. Current practice in contractor prequalification is to telephone references, particularly past clients for which the contractor has performed work, to gather his past performance data. These data are used to determine how well the contractor would be expected to perform on the project in question. In most instances, there are numerous performance-related parameters of interest (e.g. "quality of work performed," "timeliness of project completion," "accuracy and timeliness of paperwork," and others).

However, the following problems associated with this data collection technique can be identified:

FROM THE INQUIRER'S (OWNER) PERSPECTIVE:

1. Only limited amounts of data can be obtained by telephone due to time constraints and format (lack of structure) in which the data is requested and collected;

2. Usually only a small number of references are phoned, therefore, a small sample can result in a biased sample and subsequently in an incorrect decision;

3. Difficulties (memory capacity and capabilities) in synthesizing all the performance data gathered for each contractor from several references to make a rational decision.

FROM THE REFERENCE'S (PAST CLIENT) PERSPECTIVE:

1. The inability to provide immediately:

   a. accurate assessments of the numerous decision parameters which characterize past performance;

   b. verbalization of the performance in distinctive categories (e.g. superior, excellent, good, average);

2. Reluctance to be truthful and candid about a candidate contractor's performance due to legal ramification or actions that could result.

Therefore, a questionnaire can be constructed to structure the data collection process with regard to the contractor's past performance by (anonymously) sampling past clients on selected decision parameters. A portion of a typical questionnaire is presented in Figure 4. Any number of decision parameters related to a contractor's past performance can be sampled, as shown in Figure 4. Example parameters include "safety," "cooperation with owner representative(s)," "quality of work performed" and number and quantities (dollars) of "charges for change orders." Respondents are requested to rate each given parameter on a scale from "superior" to "unsatisfactory."

These collected data can be presented in the form of a histogram that represents the frequency distribution of each response category. From the histogram, categorical probabilities can be calculated. Subsequently, a discrete cumulative probability distribution function can be constructed.

To illustrate this process, example data of 25 past clients were constructed on selected decision parameters regarding "Contractor X's" past performance. One decision parameter considered was "quality of work performed." The histogram of the responses is presented in Figure 5 and the discrete cumulative probability distribution function is presented in Figure 6. This information is used to infer the anticipated contractor performance regarding the subject project.

The mean and standard deviation of the collected data can be viewed to be indicative of the future performance of the contractor associated with a given factor. From Figure 5, the mean performance value of 2.72 with a sample standard deviation of 1.37 was calculated. The 95% confidence interval for the mean performance rating is between 2.17 and 3.27, respectively. It can be concluded that the contractor's future performance of "quality of work performed" factor would be rated between "excellent" and "good."

After the completion of these steps, the decision maker must interpret the numeric results by selecting threshold values for acceptable and unacceptable performance. These values must be based on experience and engineering judgment. A decision rule can be constructed to aid in the decision process, such as the one presented below:

IF

    Factor score < 4.00

THEN

    Factor performance = Acceptable

Decision Parameter	Superior	Excellent	Good	Average	Poor	Unsatisfactory	Comments
1. Safety							
2. Quality of Work Performed							
3. Field Supervision							
4. Field Support From Main Office							
5. Job Planning and Scheduling							
6. Job Equipment and Facilities							
7. Charges for Change Orders							
8. Accuracy and Timeliness of Paperwork (i.e. Billing, etc.)							
9. Cooperation with Personnel							
10. Response to Off-Hour Requests							
11. Overall Rating							

How long has the contractor performed work under your supervision? ____ years
What line of work is this contractor primarily engaged in? _____

FIGURE 4. Typical Parameters Included in a Questionnaire

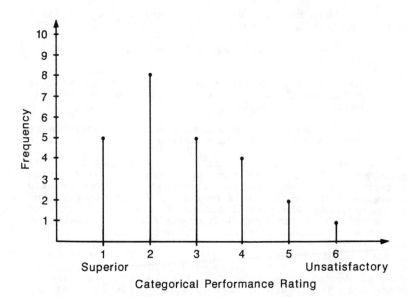

FIGURE 5. Histogram of Example Data on Decision Parameter "Quality of Work Performed"

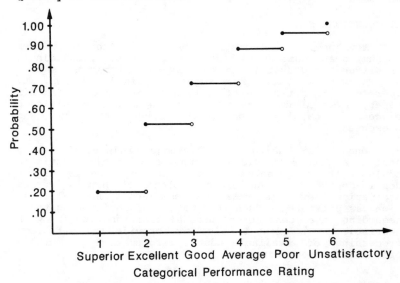

FIGURE 6. Discrete Cumulative Probability Distribution Function For Decision Parameter "Quality of Work Performed"

Based on the results of the example, the decision parameter "quality of work performed" would be acceptable.

The following shortcomings with this approach can be identified:

1. The number of observations is still small so that a 95% confidence of the mean performance value is difficult to obtain;

2. The reliability and accuracy of the questionnaire data is suspect for many reasons.

This demonstrates one application of a statistical technique with regard to the contractor prequalification process. This technique is augmented by the heuristic knowledge embedded in decision-making. Such integration of engineering analysis tools with the heuristic and judgmental rules can provide for efficient processing of available data before arriving at a conclusion.

The technique outlined above is a structured approach to determining the performance of a contractor relative to selected decision parameters. The presented process is intended to aid in making more rational decisions. At any time during decision-making, the "user" (human decision maker) may override the system.

4. CONCLUSION

The knowledge acquisition process is the most time consuming and cumbersome task in the process of building a knowledge-based system. A knowledge-based system is only as "good" as the knowledge contained within it. Therefore, this task is of crucial significance. In this paper, a structured review of the knowledge acquisition process regarding contractor prequalification was presented.

One issue facing the knowledge engineer is which analytical tools should be combined with the heuristic tools. It was concluded that this decision is highly domain-dependent. An example blend of heuristic and statistical analysis regarding contractor prequalification was presented. Such integration of engineering tools regarding the contractor prequalification domain with heuristic or judgmental rules can provide for efficient processing of available data before arriving at a prequalification decision.

APPENDIX.—REFERENCES

Conover, W. J., (1980). Practical Nonparametric Statistics, Second Edition, John Wiley & Sons, Inc., New York, NY, pp. 256-263.

Diekmann, J. E., (1981). "Cost-Plus Contractor Selection," Journal of the Technical Councils, ASCE, Vol. 107, No. TC1, April, 1981, pp. 13-25.

Grover, M. D., (1983). "A Pragmatic Knowledge Acquisition Methodology," Proceedings of the Eighth International Joint Conference of Artificial Intelligence, Karlsruhe, West Germany, pp. 436-438.

Harman, H. H., (1976). Modern Factor Analysis Third Edition, University of Chicago Press, Chicago, IL, and London, England.

Hart, A. (1986). Knowledge Acquisition For Expert Systems, McGraw-Hill Book, U.S.A.

Nguyen, V. U., (1985). "Tender Evaluation by Fuzzy Sets," Journal of Construction Engineering and Management, ASCE, Vol. 111, No. 3, September, 1985, pp. 231-243.

Russell, J. and Skibniewski, M., (1987b). "A Structured Approach to the Contractor Prequalification Process in the U.S.A.," Proceedings of the CIB-SBI Fourth International Symposium on Building Economics, Copenhagen, Denmark, September 14-17, 1987.

Russell, J. and Skibniewski, M., (1987c). "Decision Criteria in Contractor Prequalification," to appear in the Journal of Management in Engineering, ASCE, April, 1988.

Russell, J. et al., (1987d). "Decision Variables For Contractor Prequalification," submitted to the Journal of Construction Engineering and Management, ASCE, November, 1987.

Waterman, D. A. (1986). A Guide to Expert Systems, Addison-Wesley Publishing Company, U.S.A.

PART IV

GENERAL

Logic programming to manage constraint-based design

Weng-tat Chan[1] and Boyd C. Paulson, Jr.[2], M.ASCE

Abstract

Engineering design involves the evaluation and satisfaction of a wide variety of constraints where constraints represent the necessary and correct relationships among design parameters and design factors. The exploratory nature of the preliminary design process necessitates the ability to represent and process these constraints in the computer, especially if designs are to be verified. However, constraints need not only check designs but can also be used to derive design solutions that satisfy the required relationships. The dual role of design constraints can be achieved through a fundamental change in the manner design knowledge is represented and organized in the computer so as to emphasize the inter-relatedness of the design parameters. This view parallels the declarative interpretation of programs in logic programming where program statements are declarations of relationships between objects. But logic programs also have a procedural interpretation which can be used not only to verify but also to maintain the relationships to-date. The paper discusses the relation between design constraints and the Prolog programs used to implement the constraints, and the implementation of a design interpreter in Prolog that utilizes the concept of constraints. Constraint conflicts are part of the design process. We also discuss a simple scheme, involving the partitioning of the design into modules, to effect design changes when constraint conflicts occur.

1. Introduction

The purpose of design is to derive suitable values for the design parameters subject to their satisfying the restrictions imposed by the design specifications. Feasible designs not only satisfy the specifications but take into account other factors like the strength and properties of materials in which the design is to be executed, the causal mechanisms determining the function, operation and behavior of the design artifact, and also such factors as the cost and the capabilities of the manufacturing technology available (Mostow, 1985).

---

[1] Lecturer, Department of Civil Engineering, National University of Singapore Kent Ridge, Singapore 0511.

[2] Professor, Department of Civil Engineering, Stanford University, Stanford, CA 94305-4020, USA.

This implies that there is a rich web of inter-connections among the design parameters and design factors which designers have to take into consideration when proposing design solutions. These interconnections form a system of 'constraints' on the admissible values of the design parameters because assigning a value to any parameter in the system has the potential of restricting the range of values of the other parameters connected to it. Furthermore, a change to any parameter should induce corresponding changes in other parameters so that the relationships are maintained.

During preliminary design, many design decisions regarding values for design parameters are tentative in nature. Some of these decisions will have to be reconsidered when their implied consequences cannot meet later specifications or are in conflict with other decisions. In both cases, some of the constraints in the system cannot be satisfied with the current assignment of values to the design parameters and some of the design values must be changed. At this point, it would be helpful to have answers to questions like:

- what are the specifications or constraints involved,

- what were the decisions that led to the conflict,

- will a new assignment of values to the design parameters involved succeed in resolving the conflict,

- what assignment of values to design parameters will satisfy a particular constraint, and

- how will this affect other design parameters. Furthermore, any change to a design parameter should be reliably propagated to the other constraints so that the design description is properly updated.

Constraint conflicts and design changes are part of the design process. Since designers cannot always be making the correct choices the first time around, it is necessary that there be a scheme to handle constraint conflicts and design changes when they arise. Unfortunately, the predominant situation is that design knowledge is organized in design programs around the representation of the design parameters and the procedures to calculate descriptions of these parameters. This makes it difficult to answer the kind of questions above that are most often asked during preliminary design. What is required is a fundamental change in the way this knowledge is organized - from a representation focussing primarily on the parameters to one that gives equal emphasis to the verification and maintenance of relationships between the parameters. The first part of this paper will discuss how this is achieved by adopting a constraint-based representation of design knowledge, and the way constraints are implemented using Prolog (Clocksin and Mellish, 1984; Hogger, 1984), a member of a class of logic programming languages. Although any other language could have been used, it is perhaps fitting to use Prolog because logic programming introduced the idea of the declarative interpretation of computer programs which supplements the usual procedural interpretation. Prolog also has many unique features, like built-in unification and meta-level programming capabilities, which make it suitable for the kind of work reported here.

However, Prolog does have its limitations, particularly its standard control of inference and backtracking strategies which are too rigid for the requirements of the kind of exploratory design outlined above. In the second part of this paper we discuss the key elements of a model of the design process based on the verification and satisfaction of constraints that is intended to correct these deficiencies. We will also discuss the main outline of the design interpreter which uses the model of design proposed.

2. Constraints

A constraint is a relation between a set of parts. In the context of this paper, the parts are the design parameters for which descriptions have to be derived as values. The values of these parameters may be described qualitatively or quantitatively but the constraint enforces the intended relation between the descriptions of the parameters. Design and code specifications that limit the allowable values of the design parameters are an important source of constraints in engineering design. They represent the traditional passive role of constraints as design checks since they eliminate all combinations of design descriptions except those that satisfy the specifications. When all the parameters related by the constraint contain design descriptions, the constraint can be invoked to check if the relation specified by the constraint holds. For example, the inequality $d < Ds$ expresses the requirement that the design stress of a structural member, d, should be less than the allowable stress, Ds set forth in the design code and is checked only if both the member and the allowable stresses were known.

Functional dependencies between physical quantities may also be considered as constraints since they constrain the quantities involved in relation to one another. Assigning a value to one physical quantity implicitly constrains all other quantities linked to it by functional dependencies. Viewed in another way, physical quantities that are thus constrained have the ability to adjust their values in relation to one another so that the relation is maintained. For example, Ohm's Law $V=IR$, specifies the relationship between the current, I, and voltage, V, in a resistor of resistance R. If the voltage is changed, the current flowing through the resistor will adjust itself to maintain the relationship expressed by Ohm's Law; likewise the voltage changes when the curent is changed. We can of course use Ohm's Law to deduce the resistance R when the voltage and current are known. In exploratory design, we would like to capitalize on the ability of such constrained quantities to adjust themselves so as to maintain the constraint's relationship. Of course, the constraint can also be used to check on the validity of the values of the quantities constrained if all the values are known at the time of the check.

The present generation of design analysis programs organize the knowledge of functional dependencies as procedures that calculate design descriptions for design parameters. Design specifications are represented separately. This manner of representation has several short-comings:

- the programmed code can only be utilized for the purpose for which it was written. In consequence, these programs always calculate certain output parameters from a predetermined set of input parameters. They lack the flexibility needed to 'tweak' any part of the design description and have the effect of these changes propagated consistently throughout the entire design description.

- the knowledge of interdependencies between the design parameters is left implicit in the procedures which calculate the design descriptions. Design programs need to represent these dependencies explicitly in order to offer better support during the design stage when decisions are tentative and constraint conflicts are bound to occur. The same procedural knowledge for computing parameter descriptions can be organized around the verification and maintenance of the relationships between the design parameters if these relationships are represented explicitly as first class objects, rather than implicitly in programs. In the next section, we discuss why Prolog forms a suitable foundation both for the representation of constraints as well as the procedures necessary to maintain and verify constraints.

2.1 Prolog

Prolog, which stands for PROgramming in LOGic, is the most successful and popular of the logic programming languages. Developed by Colmereaur and his associates in 1972, it achieved prominence after the Japanese government identified it as the language for systems development in their Fifth Generation Computer research project. What follows is a brief description of Prolog, intended only to give a feeling of some of the unique features of the language and is not meant to be exhaustive or complete.

Prolog is based on the principle of automated theorem-proving by logical inference. Prolog programs consist of statements about the relationships between objects in the problem domain. These statements (or predicates) may be in the form of simple assertions (called facts) like

((father Tom, George));; Tom is the father of George
((mother Jill, George));; Jill is the mother of George

or as rules like

((parent $\_P$ $\_C$)
  ($\overline{father}$ $\_P$ $\_C$))

((parent $\_P$ $\_C$)
  ($\overline{mother}$ $\_P$ $\_C$))

```
;; An individual, _P, is the parent of another individual _C if _P
   is the father of _C. Alternatively, _P is the parent
   of _C if _P is the mother of _C.
```

Prolog tries to establish the admissibility (truth or falsity) of other statements presented in the manner of queries, like

```
? ((parent Tom George))

;; Is Tom the parent of George?
```

A query, which is a top-level goal, can be answered in the affirmative if its truth can be established from the facts and rules contained in Prolog's knowledge-base. Prolog uses a highly efficient form of inference based on the resolution principle (Robinson, 1965) to accomplish this. A query is answered in the negative (false) if it cannot be proved true.

Prolog combines theorem-proving with unification to provide even more powerful problem-solving capabilities. Unification, in simple terms, means making two objects identical. Since unification is built-in, the Prolog programmer is spared the details of assigning values to variables to unify two objects. Unification permits more interesting queries like

```
? ((parent _P George))

;; Who is the parent of George?
```

or even

```
? ((parent _P _C))

;; Retrieve any pair of individuals, _P and _C, who stand
   in the relation of _P being the parent of _C.
```

The latter queries demonstrate Prolog's ability to 'discover' individuals in its knowledge-base that satisfy relationships as a by-product of proving the admissibility of these relationships. Backtracking, another built-in capability of the Prolog interpreter, will discover all such individuals which satisfy the stated conditions in the goal predicate.

As the examples indicate, Prolog programs consisting of predicates, can be read declaratively as statements about the relationships between objects in the problem domain. Because of the close similarity between the concept of constraints and predicates, Prolog seems a natural choice as a language to implement constraints. We will discuss how Prolog programs can be used to implement constraints after a review of the previous uses of constraints in design.

2.2 Constraints in design

The work implemented in Thinglab (Borning 1979), a constraint-based simulation system, introduced the idea that onstraints could not only check descriptions but propose new ones that satisfy the constraints. The work by de Kleer (1978) and Stallman and Sussman (1977) used constraint-based representations to analyze and synthesize electrical circuits. Their work also introduced the idea of dependency-directed backtracking to correct contradictory confluences, i.e., different values for the same parameter proposed by different constraints.

Holtz (1982) represents design constraints as algebraic inequalities and uses algebraic simplification to derive the values of certain designable items in terms of given parameters. This approach has the advantage that the limits placed on a designable item by various constraints can be reflected in the final algebraic expression for that item. The disadvantage is that algebraic manipulation of constraint expressions often proves intractable when the causal chain gets longer and cannot handle the occurrence of confluences, i.e., when the constraint network has loops.

Maher (1984) recognized the use of constraints for ensuring compatibility between the different stages of design considered by the HI-RISE system. However, each stage of the design is completely determined before the next stage is undertaken and there is no backtracking to reconsider design choices in a previous stage.

The Edinburgh Designer System (EDS) (Popplestone 1984) is very close in spirit to the work reported here. It also uses modules to partition design parameters and constraints, where a module can either describe a part or the interactions between parts. However, it uses algebraic manipulation to propagate and solve for design values. Dependency-directed backtracking is performed by a separate assumption-based truth maintenance system.

Other closely related work is that by Rasdorf and Fenves (1986) and Rasdorf, Ulberg and Baugh (1987) on structural design constraints maintained in a relational database.

2.3 Representation of constraints

We represent constraints by procedural attachment (Sussman and Steele, 1980). Prolog procedures are written for each intended use of the constraint and associated with it so as to maintain the constraint relation when invoked. The procedures are pattern-invoked depending on the current design context. The constraint propagator, a program responsible for constraint maintenance, checks the design context and deduces the appropriate use of a constraint. The proper procedure can then be invoked to maintain the constraint. For example, the constraint $a=b+c$ would be represented as:

```
((constraint 1 a b c)
    (( 0 1 1/ 1 0 1/ 1 1 0) (SUM b c a))
    ((1 1 1) (SUM b c _X) (eqq _X a)))
```

There are two attached procedures in the constraint, the first to handle the case when it is necessary to derive the value of a constrained variable from knowledge of the other two, and the second to check that the values of the constrained variables satisfy the constraint. Each attached procedure can be read declaratively as the relationships that are true between the parameters of the constraint when the procedure is invoked. Alternatively, the program statements can be interpreted as the sequence of actions, which if successfully carried out, will either verify the relationships or calculate values for the parameters in the constraint to satisfy the relationships.

The SUM predicate is a Prolog primitive that, given any two of its three arguments, will calculate the remaining argument so that the addition relation is maintained. Alternatively, given all three arguments, SUM will check that the addition relation holds among the arguments. Although the predicate is sufficient to implement all possible uses of our constraint, the occurrence of errors due to numerical precision forces us to treat equality differently by allowing a margin of error. Many Prolog primitives support a declarative reading like SUM, but attached procedures are still necessary to cover those constraints that are not handled by Prolog primitives or where complicated expressions involve the use of primitives that are non-declarative.

Attached procedures are always prefaced by variable instantiation patterns to indicate the conditions under which an attached procedure can be invoked. The pattern of 1's and 0's denote the status of the corresponding variables declared in the head of the constraint - a '0' indicated that the variable is uninstantiated and a '1' that the variable should be instantiated when the constraint is checked. The designer is free to leave out or include any pattern to express his preference for the use of the constraint. It is not necessary to support all possible uses of a constraint with attached procedures. The tasks of pattern matching, assigning values to the constraint variables and invoking the appropriate procedures are handled by the constraint propagator.

3. A constraint-based model of design

Computing with constraints to derive design descriptions requires a computational model of the design process in terms of satisfying constraints and resolving constraint conflicts. We discuss the key elements of the prototype model developed to-date.

3.1 Constraint propagation

A constraint network is formed whenever a part, represented by a variable, is constrained by two or more relations. For example, the following set of linear equations forms a constraint network:

$a = b + c$
$b = 10c$
$b > 10$

The constraint network together with the variables constrained by the network make up the constraint system.

Computation in constraint systems is performed by a series of one step deductions called constraint propagation. It is initiated when a change to a constrained variable occurs. This 'activates' the constraints which constrain the variable and new descriptions are calculated for the other variables to preserve the affected relationships. These changes, when propagated to the rest of the constraint system, cause other constraints to be activated. Constraint propagation consists of repeatedly identifying the activated constraints, deriving new descriptions for the constrained variables and propagating the changes to the rest of the constraint system. The process ends when no new descriptions are generated and the constraint system achieves a new equilibrium. Constraint propagation implements a kind of antecedent reasoning and efficiently exploits the dependencies in the constraint system to derive a system-wide consistent solution. The dual role of constraints permits the derivation and checking of design descriptions to be integrated.

In the constraint-based exploratory design model, the search for solutions is a continual process of formulating, propagating and integrating constraints. Constraints are formulated from the initial problem specification. However, constraints may be added, deleted or changed as the design proceeds but any changes to the design or the specifications should always be evaluated against the constraints. The important task of maintaining the consistency of the design description in the face of such changes should be automated. Design descriptions are obtained and the state of the design is advanced when constraints are propagated. Constraint propagation derives all the valid consequences of the current state implied by the constraints. The process can be initiated on any part of the design, stopped to examine the partial results, and restarted either from where it was stopped or in an entirely different part of the design. Since constraints express the required relationships between the various parts of a design, the order in which constraints are used does not affect the correctness of the result. Constraint propagation discovers locally feasible solutions defined by each individual constraint. We integrate constraints when locally feasible solutions satisfy all constraints and are therefore globally feasible. A constraint conflict occurs when a local solution is not feasible globally. Design methods which use constraint propagation have to deal with the prospect of failure caused by constraint conflict and devise means of recovering from the conflict. We will examine the scheme used to resolve constraint conflicts in a later section after a discussion of the other key elements of the constraint-based model of design as the conflict resolution scheme depends on these other elements.

3.2 Heuristics

During the course of design, there would have been many decisions in arriving at the values of certain key parameters. These decisions are often based upon knowledge of other design parameters. When the implications of a decision taken at some point in the design conflicts with those of another decision, it becomes necessary to identify those decisions which led to the conflict and review some of them. The basis of the decisions might permit the derivation of alternative values for parameters implicated in the conflict which in turn might lead to the resolution of the conflict.

We therefore represent explicitly the logic behind these decisions as Prolog procedures and package these decisions into an identifiable unit called a design heuristic. Each design heuristic is responsible for a design parameter and it derives values for that parameter following the logic encoded in its procedure. An example of a design heuristic is the one shown below to return member sizes for a structural beam:

((bridge-heuristic Dh1 Cx1 F1)
((0 1) (OR ((LESS F1 0)
(findSection tension Cx1 "@h" beam))
((findSection compression Cx1 "@h" beam))

where findSection is a predicate that does a look-up of a table of beam sizes and picks a size according to whether the member is in tension or compression. A design heuristic may use the values of other design parameters in its decisions - these parameters become the basis of the heuristic procedure. A heuristic is able to return further alternative values for its design parameter upon re-evaluation and therefore represents a backtrackable choice-point within the design process. When all its possibilities are exhausted, the heuristic simply fails.

3.3 Meta-heuristics

It is insufficient for design purposes to have knowledge of the relationships between the design values and good heuristics to make initial assumptions. The process of design requires further knowledge of how to use the previous type of knowledge effectively. We term the latter kind of knowledge heuristic meta-knowledge. Like heuristics and constraints, it will be useful to represent and capture this kind of knowledge explicitly.

Whilst the heuristic offers advice on design values, the meta-heuristic concerns itself with issues of control over the direction of the design process. Good examples of such meta-level knowledge would be the ability to:

- rank candidate heuristics for evaluation;

- decide the worth of a particular heuristic i.e. whether it should be re-evaluated when the current value is implicated in a constraint violation;

- rank the possible causes of a constraint violation to better control the search direction and improve its efficiency.

Such knowledge is derived from experience with solving design problems using the representational facilities offered by the system the designer is using. The decision procedures in such knowledge involve the current state of the design i.e. the constraints solved, the design parameters derived or yet to be derived, the heuristics used and their effect, the dependencies between the design parameters derived and so forth.

Meta-heuristic design knowledge is represented in the form of if-then rules. The following clause is an example:

```
((meta_heuristic mH1)
       (solved-constraint DS1)
       (using-heuristic H1)
       !
       (freeze-heuristic H1))
```

The if part of a meta-heuristic lies before the "!" mark and consists of predicates that match against facts known about the current state of the design. The *then* part follows after the mark and consists of predicates that encode the advice. The advice is only carried out if all the predicates in the *if* part of the clause match the facts in the current design state.

In our example, the advice part causes the heuristic, H1 to be 'frozen' at its current value if a certain constraint has been solved using the value returned by the heuristic. The effect of this is to prevent the re-evaluation of the heuristic should the heuristic value be implicated in a constraint conflict.

3.4 Design artifacts

A design artifact is described by its collection of design parameters, constraints, heuristics and meta-heuristics. Since design artifacts may be described at different levels of abstraction and detail, not all of the design parameters, constraints etc. would be relevant at all times. We therefore partition the constraints, heuristics and meta-heuristics into modules, each module corresponding to a particular conceptualization of the design artifact by the designer. The modules may be arranged relative to each other to express such concepts as part-subpart and levels of abstraction.

Each module declares in its interface those parameters through which it is willing to exchange design descriptions. All other parameters not declared in the interface will be considered as local to the module. Interaction between any two modules therefore depends on agreement on the specification of the common interfacn between them. Constraint propagation first acts locally within a module and only selected results of the propagation are disseminated to other modules through the interface.

3.5 The design interpreter

User-written constraints, heuristics and meta-heuristics are high-level abstractions employed in the constraint-based model of design. They are implemented as Prolog programs which are interpreted and executed by another Prolog program known as the design interpreter. The coding of the design interpreter makes extensive use of the meta-level programming capabilities of the Prolog language. Essentially, this means that the design interpreter is acting like another Prolog interpreter, with all of the underlying Prolog interpreter's built-in facilities like unification and resolution theorem proving but customized in certain respects to better suit the task at hand. In particular, the design interpreter imposes a control and backtracking strategy (at the constraint level) that is more suitable for exploratory design.

The design interpreter works with the concept of design tasks. A design task covers all the actions necessary to create an instance of a design module - invoking heuristics, propagating constraints, backtracking when necessary. The prime goal of the design task is to satisfy all the constraints contained in its associated design module, and in so doing, derive values for all the design parameters in the module. A task may cause other tasks to be initiated as sub-tasks and there may be many tasks to be completed before the design is finished. Design tasks are similar to high-level Prolog goals but unlike them, design tasks need not be initiated and completed in a predetermined order and a design task need not be completed before another task can be initiated. Rather, a task may be temporarily suspended pending the availability of a design description to be provided from another design task. Another task may then be started which might contribute directly or indirectly to the completion of the first task. Eventually, the first task is taken up from where it was suspended. Prolog's control strategy involving depth first exploration of the goal hierarchy is too rigid for the kind of interaction described. Instead, a more flexible control strategy is required which allows tasks to be taken up and suspended depending on the circumstances. This is achieved by the use of agendas to record all unfinished tasks and determining the current task by choosing from the tasks on the agendas. The design interpreter utilizes two task agendas. The first one is for tasks that have been initiated but have not been completed. Constraint propagation, punctuated by heuristic invocation and the consultation of meta-heuristics, attempts to further the state of the tasks on this agenda towards completion. As far as possible all constraint conflicts that are encountered along the way are resolved by backtracking among the local choice-points of the task module where the conflict occurred. Tasks where all the constraints have been solved are removed from the agenda and marked as completed. However, they may re-enter the agenda at a later stage if a constraint conflict elsewhere requires a redesign action.

A task is moved from the main agenda to the redesign agenda when a constraint conflict occurs which could not be resolved by backtracking within the task module itself. Choice-points within other task modules that contributed design descriptions to the task module concerned are identified. One of these choice-points is selected and the task containing the choice-point is also entered on the redesign agenda. Backtracking within this task module returns a new description to the task module in which the conflict occurred. The new description may lead to the resolution of the conflict, otherwise other choice-points have to be tried. The design interpreter tries to remove all the tasks from the redesign agenda before continuing with tasks on the main agenda.

3.6 Redesign

Redesign is initiated when a constraint violation occurs. This violation implies that some of the values of the design variables involved in the constraint are incompatible with each other. Since the values of all design variables result directly or indirectly from the design choices made by the heuristic procedures, the violation implies that one or more of these choices must be retracted. Which choice to retract is an informed decision that depends on the circumstances of the constraint violation, but we can aid the designer by maintaining the dependencies between a design choice and the consequences of its use.

Dependencies between different design parameters are recorded and maintained during constraint propagation. We distinguish between three different types of design parameters for the purpose of maintaining dependencies:

- *postings*: these are design parameters whose values are determined outside the context of the current task and can be treated as given to the design procedure encoded by the constraints and heuristics in the module;

- heuristic *choices*: these are design parameters determined by the heuristics in the current module and are treated as tentative and can be retracted should inconsistencies arise as a consequence of their use;

- constraint *derived* parameters: these are design parameters that are derived from other parameters through constraint propagation. They are not considered for revision unless the values of the parameters from which they were derived have changed.

The postings and heuristic choices represent the choice-points within a task but with an important difference - heuristic choices are local to the task module and are easily explored by backtracking. Postings are treated as constants within the task to which they are posted. Only backtracking within the task from which they were posted out can change their value.

A constraint conflict indicates that the values of the design parameters involved in the constraint are incompatible with each other and at least one of them must be revised. The manner of locating the appropriate parameter to change depends on insight into the problem domain. However, weak ordering strategies can be used to discriminate among the parameters in the possibly large basis set:

i) Parameters returned by heuristics and postings are the logical parameters for reconsideration since they represent choice-points within a task module.

ii) Between heuristics and postings, it is better and easier to re-evaluate the heuristics which are local to the module than to question the values of postings which imply re-evaluating heuristics in other modules, causing design changes to occur in the other modules. This re-evaluation favors a local redesign strategy over a global redesign strategy.

iii) It is better to reconsider heuristically derived parameters in the reverse order in which they were derived, i.e., the latest heuristic in the basis is re-evaluated before heuristics that were used earlier.

Determining which parameter to change therefore translates to determining an appropriate choice-point to backtrack to. The dependency graph records the dependencies between design parameters and the choice-points of a task module. They are used in the dependency-directed backtracking strategy of the design interpreter which replaces the naive default chronological backtracking of the underlying Prolog interpreter.

The weak and uninformed strategies outlined above impose a default ordering on the basis of parameters associated with a constraint conflict. More specific domain knowledge in the form of meta-heuristics can affect this ordering by either removing certain entries in the basis or causing a partial reordering of the basis.

3.7 Micro-computer implementation

The work described in this paper is part of a system to do constraint-based design. The system is written in micro-Prolog (McCabe, Clark and Steel 1985) and runs on an IBM PC/AT with 640K of memory. Extensive use is made of the module facility of micro-Prolog to partition the software in the system into sets of functionally related predicates and clauses. These modules only interact with each other via a declared interface consisting of named predicates and constants.

One of the main limitations in micro-computer based systems is the limited amount of addressable physical memory. It was fortunate that the micro-Prolog interpreter had a program partitioning capability (using modules) and an error-trap handler which signalled the absence of a predicate when called. It was possible to implement a simple but effective paging system utilizing these two features which moved the program modules between physical memory and a virtual disk depending on the requirements.

4. Conclusion

The relationships between design parameters need to be explicitly represented if design programs are to provide better support of the kinds of decision processes occurring during the preliminary design phase. The procedural knowledge currently organized around the derivation of parameter descriptions can instead be organized around the maintenance and verification of constraints.

Logic programming languages form a good basis for the representation of constraints as well as the implementation of the procedures to maintain and verify constraints because of the declarative and procedural interpretations of logic programs.

In Prolog, the imposition of a standard control and backtracking strategy in the program interpreter necessitates the implementation of a customizable design interpreter. However, Prolog's meta-logical programming capability and the ease with which complex systems can be rapidly prototyped, greatly facilitate this task.

References

Borning, A. 1979. "ThingLab - A Constraint-oriented Simulation Laboratory", Ph.D thesis, Dept. of Computer Science, Stanford University, California.

Chan, W. T. 1986. "Logic Programming for Managing Constraint-Based Engineering Design", Ph.D thesis, Dept. of Civil Engineering, Stanford University, California.

Clocksin, W.F., and Mellish, C.S. 1984. *Programming in Prolog*, 2nd ed., Springer-Verlag, Berlin.

Colmerauer, A., Kanoui, H., Pasero, R. and Roussel, P. 1973. "Un Systeme de Communication Homme-machine en Francais", Research Report, Groupe Intelligence-Artificielle, Universite Aix-Marseille II.

de Kleer, J., and Sussman, G. J. 1978. "Propagation of Constraints Applied to Circuit Synthesis", M.I.T. Artificial Laboratory Memo 485, Cambridge, Massachusetts.

Hogger, C. J. 1984. *Introduction to Logic Programming*, APIC Studies in Data Processing No.21, Academic Press Inc. (London) Ltd.

Holtz, N. M. 1982. "Symbolic Manipulation of Design Constraints - an Aid to Consistency Management", Ph.D. dissertation, Dept. of Civil Engineering, Carnegie-Mellon University, Pittsburgh, Pennsylvania.

Maher, M. L. 1984. "HI-RISE: An Expert System for the Preliminary Structural Design of High Rise Buildings", Ph.D. dissertation, Dept. of Civil Engineering, Carnegie-Mellon University, Pittsburgh, Pennsylvania.

McCabe, F. G., Clark, K. L., and Steel, B. D. 1985. *Micro-Prolog Professional 1.2 Programmer's Reference Manual*, Logic Programming Associates, Ltd., London, England.

Mostow, J. 1985. "Toward Better Models of the Design Process", *AI Magazine*, Spring, 44-57.

Popplestone, R. J. 1984. "The Application of Artificial Intelligence Techniques to Design Systems", *International Symposium on Design and Synthesis*, Japan Society of Precision Engineering, Tokyo.

Rasdorf, W. J., Ulberg, K. J., and Baugh, J. W. 1987. "A Structure-Based Model of Semantic Integrity Constraints for Relational Databases", *Engineering with Computers*, Springer-Verlag, vol. 2, no. 1, Spring, 31-39.

Rasdorf, W. J., and Fenves, S. J. 1986. "Constraint Enforcement in Structural Design Databases", *Journal of the Structural Division*, American Society of Civil Engineers, vol. 112, no. 12, December, 2565-2577.

Robinson, J.A. 1965. "A Machine Oriented Logic based on the Resolution principle", *Journal of the ACM*, vol. 12, 25-41.

Stallman, R. M., and Sussman, G. J. 1977. "Forward Reasoning and Dependency-directed Backtracking in a System for Computer-aided Circuit Analysis", *Artificial Intelligence*, vol. 9, 135-196.

Sussman, J. G., and Steele, G. L., Jr. 1980. "CONSTRAINTS - A Language for expressing Almost-Hierarchical Descriptions", *Artificial Intelligence*, vol.14, no.1, August, 1-39.

EXPERT SYSTEMS FOR THE EARTHQUAKE-RELATED INDUSTRY

Jongeup Kim[1], Weimin Dong[2], Felix Wong[3], Haresh C. Shah[4]

ABSTRACT

This paper summarizes the development of Insurance and Investment Risk Analysis Systems (IRAS) that provides consultation for the earthquake insurance and investment banking industries. The feature of this system will be briefly described in this paper, including: interactive input/output facilities, graphic data retrieval, hierarchical knowledge-based management, integrated system of independent program modules, combination of backward-chaining and forward-chaining inference mechanisms, and approximate reasoning scheme based on fuzzy set theory to deal with linguistic and/or incomplete information.

1. INTRODUCTION

The potential financial and operational impact of a destructive earthquake on insurers and real estate investment bankers can be considerable. A major concern of the professionals working in these industries is their ability to estimate the structural vulnerability and risk under future catastrophic earthquakes. There are various questions that they need to answer for their decision-making. Some of these questions are:

1) Is a particular building, or group of buildings, in a given region insurable? What is their vulnerability? What will be the probable maximum loss (PML) of these buildings?
2) What will be the real estate investment portfolio risk for a given region due to a catastrophic earthquake?
3) Should an investment on a particular building be made from the earthquake vulnerability and risk point of view?
4) What should be the level of premium that a company should charge for different regions with varying seismic hazards?

These and other such questions can only be answered if expertise from seismology, geology, structural and geotechnical engineering, econimcs, etc., is available to the decision maker.

[1] Graduate student, Dept. of Civil Engineering, Stanford University, Stanford, CA 94305-4020.
[2] Acting Associate Professor, Dept. of Civil Engineering, Stanford University, Stanford, CA 94305-4020.
[3] Senior Associate, Weidlinger Associates, Palo Alto, CA 94304.
[4] Professor and Chairman, Dept. of Civil Engineering, Stanford University, Stanford, CA 94305-4020.

Recent advances in earthquake engineering have permitted experts to rationally estimate seismic effects, but this expertise is not readily available to managers in insurance and banking industries. There are three major problems for these managers:

1) What are the relevant data for seismic risk evaluation?
2) Where to obtain these data?
3) How to use these data for their decision making?

This paper addresses the development of expert systems called IRAS (Insurance and Investment Risk Analysis Systems), that provides consultation for the earthquake insurance and investment banking industries in California. IRAS has the most complete data base for seismic risk evaluation. Its features include interactive input/output facilities, graphic data retrieval, hierarchical knowledge-based management, integrated system of independent program modules, combination of backward-chaining (goal-driven) and forward-chaining inference mechanisms, and an approximate reasoning scheme based on fuzzy set theory to deal with linguistic and/or incomplete information and inherent uncertainty. These features will be described briefly in the following sections. The complete system runs on an IBM AT personal computer and is written in FORTRAN, a most popular programming language in scientific computation.

2. DOMAIN KNOWLEDGE

Seismic risk is defined as the likelihood of loss due to earthquakes and involves four basic components: hazards, exposure, vulnerability and location. These factors are further defined below (Miyasato et al. 1986):

a. The hazards or dangerous situations may be classified as follows:

  1) Primary hazards (fault break, ground vibration);
  2) Secondary hazards, which are triggered by primary hazards and which are potentially dangerous, e.g., a fault break causing a tsunami or ground shaking resulting in foundation settlement, foundation failure, liquefaction, landslides, etc.;
  3) Tertiary hazards, which are produced by flooding, by dam break, and by fire following an earthquake and the like.

  All these hazards lead to damages and losses. They may be expressed in terms of severity, frequency, and location.

b. The exposure is defined as the value of the structures and contents, business interruption, lives, etc.

c. The vulnerability is defined as the sensitivity of the exposure to the hazard(s) and the location relative to the hazard(s).

d. The location is defined as the position of the exposure relative to the hazard.

Losses resulting from seismic hazard are numerous and can be categorized as follows:

* Life and injury.
* Property.
* Business interruption.
* Lost opportunities.
* Tax base.
* Other losses.

A seismic risk analysis requires the identification of the losses to be studied as well as the identification of the hazard exposures and their locations and vulnerability.

For the insurance and real estate industry, property losses are the major concern. Property loss is usually measured by the damage ratio, which is defined as the repair cost of the damaged facility divided by the initial cost of the facility. Due to uncertainties in predicting structural behavior during future earthquakes, the current practice of the insurance industry in California is to use PML (probable maximum loss) as the basis for premium calculation (Steinbrugge et al. 1982). PML is defined as the damage ratio in such a way that during the "maximum probable" earthquake, 9 out of 10 buildings will experience damage less than the value given by PML.

PML does not consider the randomness of earthquake occurrence with respect to the time, place, and size of earthquake (Shah et al. 1984; Chiang et al. 1984; ATC 1986). In order to reflect the uncertain nature of earthquake occurrence, a second index, called damage threshold (DT), is used which combines the uncertain response of the building with the random occurrence of future earthquakes. Both indices are used in IRAS.

Corresponding to the major components of seismic risk, IRAS is divided into three subsystems: SHES, SRES-1 and SRES-2. The seismic hazard evaluation system (SHES) combines hazard and location components to obtain the seismic hazard estimation. The main flow chart for SHES is shown in Fig. 1. SRES1, the seismic risk evaluation system, is used to screen the property loss from exposure and vulnerability of the building. In this level, only building type (classification) is required. The flow chart for SRES-1 is shown in Fig. 2. Fig. 3 shows the flow chart of SRES-2, which performs the second level of seismic risk evaluation taking specific information of the buildings into consideration. The data management and inference mechanism of these subsystems will be described in the following sections.

3. INFERENCE MECHANISM (INFERENCE ENGINE)

An inference engine incorporates reasoning methods which act upon input data and knowledge from knowledge base to solve the desired problem and produce an explanation when requested. The control strategy for an inference engine could be forward-chaining,

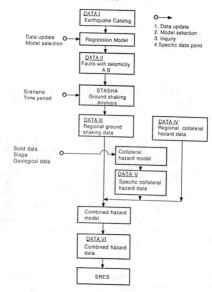

Figure 1

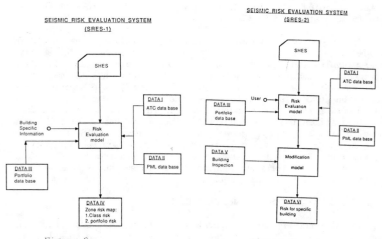

Figure 2

Figure 3

backward-chaining or a mixture of both.

In the IRAS application, the systems should be able to satisfy diverse goals (inquiries) such as "what is the real estate investment portfolio risk for a given region due to a catastrophic earthquake?", "what is the probable maximum loss of a particular building due to all contributing fault seismicity?". The goal specifies the reasoning path that should be pursued. Hence, it is natural that backward chaining (goal-driven) should be used. However, when the goal is specified and the reasoning path to achieve this goal is identified, the systems will use forward chaining (data-driven) to collect the relevant data, either by querying the user or by searching and retrieving it from the knowledge base. Thus, the control mechanism is a combination of backward chaining and forward-chaining. Since goal specification significantly reduces the search space, only a minimum search effort is required.

4. KNOWLEDGE (DATA) BASE

The knowledge (data) base for IRAS consists of raw data, production rules, engineering and analysis programs, and approximate reasoning schemes. Unlike conventional rule-based systems which use If-Then rules only, IRAS recognizes the fact that seismic risk evaluation needs both judgmental expertise and well-established mathematical procedures. Hence, IRAS incorporates both If-Then production rules and algorithmic programs. For instance, model selection depends heavily on the expert's subjective judgement, and If-Then rules are suitable to guide the user to select the appropriate model. After the model is selected, the relevant procedures are executed using algorithmic programs.

Combining inference rules with algorithmic programs is also necessary for the following reason: in most cases during inference, when the facts match the antecedents of a particular rule, the rule is triggered and the consequent can be retrieved directly from the knowledge base without further computing. When the conditions do not match the antecedents of any rule in the knowledge base, the systems will refer to the relevant programs to calculate the consequents (results).

This approach saves a great deal of computation, a consideration especially important for micro-computer implementation. Obviously, it is applicable only for problems where the inference mechanism is well-defined (as regular computational programs). For loosely structural inference mechanisms, the partial matching problem is resolved through default (applying prior information). In this case, the reliability of the consequence is reduced.

5. INTEGRATING INDEPENDENT PROGRAMS

A common practice in programming is to have a main driver and many subroutines. The driver and subroutines are compiled into a global executable program. However, when the problem to be solved is complex, many subroutines and submodules are needed. The size of the program increases rapidly and soon the capacity of the internal memory of a microcomputer is exceeded. It is then necessary to rely on fancy input-output manipulation and peripheral storage to fit the program into the computer. Furthermore, when any submodule of the program needs to be changed due to technology or engineering advances, the relevant routines must be changed and recompiled. The fitting must be reconstructed.

To facilitate upgrading IRAS and to ease the restriction of internal memory on the IBM AT, the submodules of the systems are written as independent programs. Most of these progrms have already been developed during the past ten years by staff and students at the Blume Earthquake Engineering Center, and they are simply ported to the microcomputer. The programs are compiled independently and invidually, and are then called into memory when needed by the driver much like a subroutine. The retrival, execution, and then return of the external programs is easily achieved on the IBM AT using the interrupt feature of DOS. Each external program can be as large as the total internal memory of the AT (currently at 640K byte).

6. UNCERTAINTY TREATMENT

As mentioned in the previous sections, there are uncertainties involved at each stage of the evaluation process. Earthquake occurrence is random in nature, and so is its size. For this type of uncertainty, the probabilistic approach has been well established and the data in California is good enough to use this approach. Hence, IRAS adopts the current probabilistic approach for hazard analysis to handle uncertainties in predicting ground shaking for the site. The program STASHA, developed at Stanford University for hazard analysis, was incorporated into IRAS using the approach described in the previous section.

There is yet another type of uncertainty in the evaluation which cannot be handled using probabilistic methods. In evaluating the vulnerability of a building, design detail and construction quality will affect the performance of the building. The degree of damage will vary widely depending on the quality of engineering design. All these factors will significantly influence the building performance during earthquakes and must be identified in order to get a reasonable evaluation. When the user fails to answer the inquiry on these factors from the systems, then it is expected that the system will give an answer with a wider spread due to the larger uncertainty. Because data regarding damage from diverse building types is scarce and is not sufficient to support a probability distribution, IRAS uses an uncertainty model based on fuzzy set theory to reflect the judgmental knowledge of the effect of different factors on building damage.

Fuzzy sets with different membership functions are used to represent the prior information on these effects. Some examples are given in Figure 4.

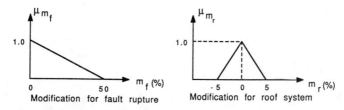

Fig. 4: Fuzzy representations for modification factors

Whenever the response to a query for data is unknown (the default), the system will use the fuzzy set instead of a crisp number to count its effect. The Vertex method (Dong et al. 1987; Dong et al. to appear) is used to combine all these effects and to calculate the total effect, resulting in a certainty factor which reflects the degree of uncertainty. When the system gives the evaluation result, it also indicates the reliability of the result (certainty factor) and how the reliability can be improved.

7. INPUT/OUTPUT FACILITIES

IRAS adopts as the main I/O facility the commercial software I/O PRO Screen Development System and Utilities V 1.1 (1985), developed by MEF Environmental. I/O PRO is a modular set of software development tools and utilities which together create a high productivity environment for FORTRAN, C, Pascal programmers. The screen development system facilities the creation of text and graphic screens used as the input and output media for interactive programs. The slides displayed on screen are used to communicate with the user for input and output and also for explanations if requested. Input data formats can be numerical, linguistic or graphical, depending upon the context. Figures 5 and 6 are examples of the slides.

Besides the interactive mode, the user can also choose the batch mode in which all data are read in together using the format of lotus 1-2-3 spread sheet or dbase III etc. This mode facilitates the data transfer from the data base in insurance and investing banking company.

In order to display the regional risk, IRAS also incorporates another commerical software, Atlas-AMP (developed by Strategic Locations Planning in 1985), to show the thematic map of regional risk. All I/O options are built into the master program and can be exercised according to the user's goal and decision needs.

210 KNOWLEDGE-BASED EXPERT SYSTEMS

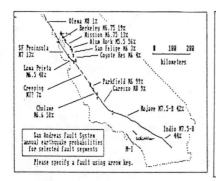 

Figure 5: An Example of Graphical Input and Output

```
┌─────────────────────────────┐    ┌─────────────────────────────────────┐
│    MAIN MENU                │    │ Condition of mechanical and electrical systems │
│    Select Risk Category     │    │ - - - - - - - - - - - - - - - - - - │
│                             │    │                                     │
│    Individual  PML  ☐       │    │          Very   good-----1          │
│    Portfolio   PML  ☐       │    │          Good--------2              │
│    Regional    PML  ☐       │    │          Average------3             │
│                             │    │          Poor---------4             │
│    Individual  SDT  ☐       │    │          Dangerous----5             │
│    Portfolio   SDT  ☐       │    │                                     │
│    Regional    SDT  ☐       │    │       Unknown  =  default--0        │
│                             │    │                                     │
│       Return to DOS  ☐      │    │          Enter number  ☐            │
└─────────────────────────────┘    └─────────────────────────────────────┘
```

Figure 6: An Example of Numerical & Linguistic Input

8. CONCLUSION

This paper summarizes the development of the seismic risk analysis systems, IRAS, for the insurance and real estate investment industry. The main framework of the systems has been completed and demonstrated to the clients and users, and the response from potential users has been excellent.

IRAS has been developed for the buildings located in California, due to its sponsors' main interest. However, in view of the modularity and flexibility of its design, IRAS can be readily adapted to other regions of the world when the appropriate data/knowledge bases are available.

REFERENCES

Applied Technology Council, Earthquake Damage Evaluation Data (ATC-13). (Redwood City, CA, 1986).

Chiang, W.L. et al, Computer Programs for Seismic Hazard Analysis, Report No. 62, The John A. Blume Earthquake Engineering Center, Stanford University (1984).

Dong, W. and Wong, S., Fuzzy Weighted Averages and Implementation of the Extension Principle, Fuzzy Sets and Systems (1987), Vol. 21, pp. 183-199.

Dong, W. and Shah, H., Vertex Method for Computing Function of Fuzzy Variables, in print.

Miyasato, G., Dong, W., Levitt, R., Boissonnade, A., Implementation of a Knowledge Based Seismic Risk Evaluation System on Microcomputer, Artificial Intelligence in Engineering, (1986) Vol. 1, No. 1, pp.29-35.

Steinbrugge, Karl V., Earthquakes, Volcanoes, and Tsunami--An Anatomy of Hazards (Skandia America Group, New York 1982).

Shah, H.C. and Dong, W., A Reevaluation of the Current Seismic Hazard Assessment Methodologies, in proc. of the 8th World Conference of Earth Engineering.

SUBJECT INDEX
Page number refers to first page of paper.

Activated sludge process, 127
Artificial intelligence, 16, 40, 127, 154
Assessments, 16

Bridges, 26
Building codes, 73
Building design, 53

Claims, 154
Computer applications, 188
Computer graphics, 26, 40, 79
Computer programming, 188
Computer programs, 118
Concrete, reinforced, 16
Constraints, 188
Construction, 140, 154, 169
Contractors, 169
Contracts, 154, 169
Control, 127
Critical path method, 140

Damage assessment, 2
Dams, 118
Decision making, 169
Design, 188

Earthquake damage, 203
Estimating, 73
Expert systems, 2, 16, 26, 40, 53, 73, 79, 88, 102, 118, 127, 140, 154, 203

Fatigue, 26
Fractures, 26
Framed structures, 16
Fuzzy sets, 16, 203

Hazardous waste sites, 102

Industrial plants, 53
Insurance, 203

Knowledge-based systems, 2, 26, 40, 53, 73, 88, 102, 118, 127, 140, 154, 169, 203

Leakage, 118

Microcomputers, 118, 188

Preliminary design, 88
Project management, 140

Qualifications, 169

Reliability, 16
Risk analysis, 203
Roofs, 73

Scheduling, 140
Seepage, 118
Seismic hazard, 203
Site evaluation, 102
Sludge treatment, 127
Snow loads, 73
Spreadsheets, 79
Steel beams, 79
Steel structures, 40
Structural design, 40, 53, 79, 88
Structural safety, 2
Structures, 2

Tall buildings, 88

Waste disposal, 102
Waste management, 102.

AUTHOR INDEX
Page number refers to first page of paper.

Adams, Kimberley, 154
Adeli, H., 40
Arulmoli, K., 118
Asgian, M. I., 118

Bedard, C., 73

Chan, Weng-tat, 188
Chen, Ruijin, 16
Connor, Jerome, 26

De La Garza, Jesus M., 140
Dong, Weimin, 203

East, E. William, 140

Fang, Hsai-Yang, 102
Fazio, P., 73

Gowri, K., 73

Ibbs, C. William, 140

Jayachandran, P., 88

Kim, Jongeup, 203
Kim, Moonja Park, 154

Kumar, B., 53

Liu, Xila, 16

Malasri, Siripong, 79
Mikroudis, George K., 102
Miller, W. O., 118

Paek, Y., 40
Parker, D. G., 127
Parker, S. C., 127
Paulson, Boyd C., Jr., 188

Roddis, W. M. Kim, 26
Russell, Jeffrey S., 169

Sanjeevan, K., 118
Shah, Haresh C., 203
Skibniewski, Miroslaw J., 169

Topping, B. H. V., 53
Tsapatsaris, N., 88

Wong, Felix, 203

Yao, J. T. P., 2

Zhang, X. J., 2